AMERICAN
CATCH

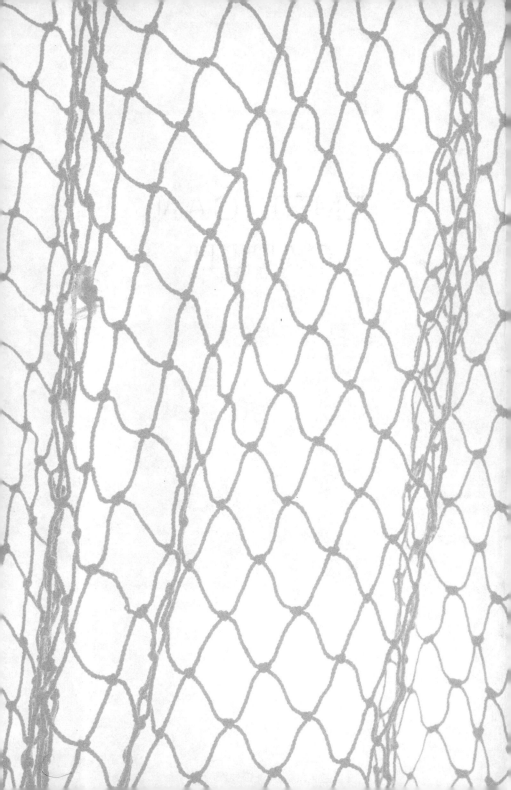

AMERICAN CATCH

THE FIGHT FOR OUR LOCAL SEAFOOD

Paul Greenberg

THE PENGUIN PRESS

New York

2014

THE PENGUIN PRESS
Published by the Penguin Group
Penguin Group (USA) LLC
375 Hudson Street
New York, New York 10014

USA • Canada • UK • Ireland • Australia
New Zealand • India • South Africa • China

penguin.com
A Penguin Random House Company

First published by The Penguin Press,
a member of Penguin Group (USA) LLC, 2014

ISBN 978-1-59420-448-7

Printed in the United States of America
1   3   5   7   9   10   8   6   4   2

Designed by Gretchen Achilles

For Dad and Sharon, real swells,

and

for Esther, deeper still

QUESTION: How do you feel about the fact that 91 percent of America's seafood is coming from abroad?

ANSWER: Who's the broad?

—*Interview with Herb Slavin,*
Fulton Fish Market fishmonger

Contents

AMERICAN CATCH

Introduction

It is a particularly American contradiction that the thing we should be eating most is the thing most absent from our plates.

Fish and shellfish are today widely recognized by physicians as central to our physical and mental health, and just about every contemporary diet—Paleo, Mediterranean, Atkins, South Beach (take your pick)—recommends seafood as a key animal protein. Heart disease, Alzheimer's, depression, even low sperm count are all conditions that a fish-based diet may help ameliorate. And by all rights this most healthy of foods should be an American mainstay. The United States controls more ocean than any other country on earth. Our seafood-producing territory covers 2.8 billion acres, more than twice as much real estate as we have set aside for landfood.

But in spite of our billions of acres of ocean, our 94,000 miles of coast, our 3.5 million miles of rivers, a full 91 percent of the seafood Americans eat comes from abroad.

Set against the backdrop of the larger American food

system, the seafood deficit, is, well, fishy. Many of our most important landfoods are trending in the opposite direction. Corn, anybody? Plenty of it—surpluses of it, in fact. Beef? Enough domestic production to supply every American with around eighty pounds a year—five times the national per capita rate of seafood consumption. Meanwhile, the paucity of domestic fish and shellfish in our markets and in our diets continues even as foreign seafood floods in at a tremendous rate. In the last half century American seafood imports have increased by a staggering 1,476 percent.

It gets fishier still. While 91 percent of the seafood Americans eat is foreign, a third of the seafood Americans *catch* gets sold to foreigners. By and large the fish and shellfish we are sending abroad are wild while the seafood we are importing is very often farmed. Two hundred million pounds of wild Alaska salmon, a half billion pounds of pollock, cod, and other fish-and-chips-type species, a half billion pounds of squid, scallops, lobsters, and other shellfish is, every year, being sent abroad, more and more often to Asia; untold tons of omega-3-rich seafood are leaving our shores to help other countries lower their rates of heart disease, raise their cognitive abilities, and lengthen their life expectancy. American consumers suffer from a deficit of American fish, but someone out there somewhere is eating our lunch.

How did we land ourselves in such a confoundingly American catch?

I first began contemplating this question one morning in the summer of 2005. I had recently moved to the far southern end of the island of Manhattan to what the seafaring Dutch had called New Amsterdam but which is now more sterilely christened the Financial District. This lifeless agglomeration of concrete and glass was not at first glance an ideal neighborhood for a writer like me whose subjects are nature and fishing and seafood. But I liked the fact that just on the other side of the skyscrapers the water lay in three directions. I was intrigued that the place where I now made my home was America's original seaport, the very spot where some of the first American seafood had been landed.

I quickly realized I was deluding myself. For a week I followed the routes that most people traveled in my new neighborhood—up Broadway, past banks and brokerage firms, past the bare asphalt slab called Zuccotti Park. On the site of the New Amsterdam colony, there was now a depressing convergence of fast-food franchises and an oddly duplicative march of bad shoe stores. I stared into the empty pit at Ground Zero and tried uselessly to find some connection to the place like a mariner throwing out an anchor into muddy bottom. The tines found no purchase.

Finally, at the beginning of the second week in my new

home, I rose early and set out by bicycle to get a broader feel for this empty-feeling place. Choosing the rising sun as my direction, I turned east, against traffic, at the street that in New Amsterdam days had been called De Maagde Paatje— the Maiden's Path—so named because it was the chosen route for Dutch girls bringing their washing down to the East River on laundry days. Today called Maiden Lane, it brought me to Pearl Street—once Parelstraat—an appellation that derives from the time when the road was paved with native oystershell. As I coasted along Pearl, I skirted over landfill that had long ago buried a swatch of salt marsh known in colonial times as Beekman's Swamp. At Dover Street, I headed east again until I arrived at Water Street. The street once marked the water's edge, but now after landfill had widened the city considerably Water Street was several blocks from the water.

But just past Water the feel of the neighborhood started to change and the ghost of a former incarnation started to reveal itself. At the corner of Dover I came across a red clapboard building that I later learned was one of the oldest functioning public houses in New York. Beyond this distinctly maritime-looking tavern stood a row of crotchety old buildings, all from the early 1800s. Hazy stenciling could be made out on a few: "Joshua Atkins & Co. Shipchandlers," "Joseph C. LaRocca & Sons Shellfish & Seafood."

From Dover I went east to Front Street and then tacked south to Peck Slip, a grand plaza in disrepair that petered away and dumped me out onto South Street.

There before me stood a metal warehouse built during the Great Depression. Weeping with water stains, it bore a straightforward, working person's declaration of purpose on its facade:

FULTON FISH MARKET · CITY OF NEW YORK
DEPARTMENT OF MARKETS

I had happened upon what had been the most storied and voluminous seafood market in the world and nobody seemed to care that it still existed.

Once upon a time, before the island of Manhattan had any physical connection to the mainland, the Fulton Fish Market was the primary point of entry for nearly every piece of seafood New Yorkers ate. Clunky barges, flat-bottomed oyster skiffs, broad-beamed Hudson River sloops all made their way to this weigh station and docked at buildings that were inherently more geared toward the sea than to the land. There, all of this wild product was off-loaded and transformed into forms the public would recognize as usable food for the larder. Oyster shuckers by the dozens queued up in the stretch that is now covered by the FDR Drive.

Prehistoric-looking sturgeon, five feet long and armored with weird cartilaginous shields, were stacked like cord-wood in the hulls of fishing boats from as far away as Kingston and sold in pearly white chunks as "Albany Beef." Shad-cutters with the particular skill to convert an intrinsically gristly mess into a smooth boneless fillet worked in the early-morning hours and laid out their product alongside the pinkish sacks of roe, smoked or raw depending on the taste of the client at hand.

New York's fishing past is not as distant as one might think. As recently as 1929, an edition of the *Fishing Gazette* reported that a hundred fishermen were registered to live in "the village" of Canarsie in what has now been swallowed whole by the borough of Brooklyn. They all sold at Fulton. In that same report it was noted that Jamaica Bay, where Kennedy Airport now sits, was a "prolific fishing ground for scallops and many terrapin were found along its marshy shores." Up until the 1920s, enough lobsters and shellfish were taken in Gravesend Bay off Staten Island to satisfy much of the city's needs. In all, one hundred wholesale fish dealers worked Fulton and sold tens of millions of pounds of fish a year.

The Fulton Market that I found in the year 2005 was of course a different place. Most of that local product was gone, yet it was still a market. Forklifts unloaded boxes of blue crab hauled up from the Chesapeake Bay, their snapping claws and gyrating finlets poking through the crate slats. Elsewhere the new-moon crescent tails of the great pelagic wanderers—the swordfish and the tunas of the open ocean—jutted out over the edges of the cleaning tables. Restaurateurs kibitzed with the dealers jockeying for entire pallets of salmon while the occasional solitary elderly gentleman sidled up to a monger he had long known and came away with a single fillet of sole.

But on the fringes of the market in 2005, another kind of transaction was starting to rule the day. Here and there a few of the old buildings that formerly had housed smokehouses and salters and chandleries were getting a makeover. The newly mullioned window sashes and smooth-poured concrete stoops spoke of an occupation forthcoming that had nothing to do with the price of summer flounder. This was good waterfront real estate with all the old-timey trappings that make gentrifiers happy. Looked at from a developer's angle, the Fulton Fish Market, with its blood, guts, and racket, stuck out like a smelly thumb. It had become despised and distrusted by the police and the health department alike. Every city ordinance it could violate it had violated. If you didn't have a feel for fish, the whole thing seemed a terrible mess. No wonder the city wanted it gone.

But I didn't want it gone. As I looked across the melee,

my heart rose. Maybe my new neighborhood wasn't so bad after all.

My next visit to Fulton was part of a larger reconnaissance mission, a circumnavigation of the entire island of Manhattan. Setting out north, I rode up along the new Hudson River Park, a park that owes its very existence to fish. A multilane highway called Westway would have been built in the park's place had scientists not discovered striped bass juveniles in the rotting piers of the West Side and demanded the then recently formed Environmental Protection Agency carry out one of the nation's very first environmental impact statements, which would eventually halt Westway in its tracks. At the Hudson River Park's terminus I pushed my bike through the untended trails of Inwood Hill Park, and then picked up the route again through Manhattan's last remaining salt marsh. I then headed down the East River, excited to take another look at the Fulton Market at the end of my journey. But when I finally ducked under the triple necklaces of the Williamsburg, Manhattan, and Brooklyn Bridges, there was nothing to be found—no more banter, no more scales and guts sloughing off into the gutters. The Fulton Fish Market was gone.

In the time since I'd last been to Fulton, the city had finalized its deal with the fishmongers and ordered the market's full-scale relocation to a far distant outpost in the Bronx. The island of Manhattan, which had once been home to dozens of waterfront markets, now had none at all. As part of the deal, all the fishmongers were required to sign a new market lease that included a pledge that they would never return to these premises and sell fish, ever again.

Few New Yorkers were upset by Fulton's departure. It seemed unsurprising that an old, dirty market had been pushed out by white-collar commerce, much in the same way that French city fathers ripped the old Les Halles market from the heart of Paris. But looked at more closely, Fulton's departure can be seen as a sign of a much bigger shift. In fact, it marked the end of a twenty-year transition beginning roughly in the early 1980s during which fish markets and individual fishmongers went from controlling 65 percent of the seafood trade to holding on to just 11 percent. Supermarkets, meanwhile, went from selling 16 percent of our seafood to selling 86 percent. Not coincidentally, it was during this same time period that the United States confirmed its status as a seafood debtor nation.

In 2005, the year of Fulton's closure, seafood imports topped five billion pounds for the first time—double what we had imported two decades earlier. All the while as foreign seafood poured into our country, our American-born fish and shellfish were leaking out the back door. From 1985 to 2005, the same period during which our seafood imports doubled, our seafood *exports* more than quadrupled. Increasingly, what Americans eat from the sea has less and less to do with their own shores.

But the United States remains a nation of coasts, of oysters and shrimp and salmon and halibut the size of barn doors and bluefin tuna that swim faster than battleships. A nation where nearly half the population chooses to live less than ten miles from the sea. So what is keeping us from eating from our local waters?

The answer lies in an intricate interplay of ecology, economics, politics, and taste. In one sense it is a story that resembles many in our country during the last quarter century, a story of a self-inflicted destruction of domestic production followed by a reckless and giddy outsourcing to Asia. But with seafood we are talking about a delete-and-replace of something infinitely more precious and ultimately irrecoverable. With seafood we are talking about the destruction and outsourcing of the very ecology that underpins the health of our coasts and our bodies.

This process reveals itself most clearly through three iconic American local seafoods: Eastern oysters, Gulf shrimp, and Alaska salmon. Each fishery is representative of a specific American seafood era, and together they offer a view into the mistakes of our past, the complications of our present, and the hopes for our future.

The fate of the Eastern oyster chronicles the destruction of one of our greatest wild foods and one of the key pieces of biological infrastructure that allowed America's seafood abundance to exist in the first place. Up until the end of the nineteenth century, oysters were a ubiquitous presence along the East Coast, both on our menus and in our estuaries. They created a taste for the local ocean and a local habitat for the fish that swam at our doorstep. But beginning in the early twentieth century, the interdependence of Americans and oysters started to fail. This was particularly notable in what was once the heartland of American oysters: New York City. In just a few decades New York went from a place where New Yorkers consumed nearly all of their oysters from local waters to a place where local oysters were associated with the worst kinds of food-borne illnesses. Once oysters ceased to be a local food source, New York City waters became a pollution free-for-all, with effluent continuing to pour into its bays and inlets largely unchecked until the Clean Water Act was passed in 1972. Today, despite measurably cleaner waters, the Eastern oyster remains severely depleted and the country's entire oyster industry hovers at a mere 14 percent of what it had been at its peak

in the early 1900s. This despite the best efforts of conservationists and oystermen committed to rebuilding the lost
American oyster kingdom. The challenges oyster restorers
face today prove just how difficult it is to rebuild a seafood
system that has been destroyed, and impresses upon us the
imperative to preserve the intact seafood systems that we
still have.

Gulf shrimp reveal the quandaries of our immediate
seafood present and the complexities of the modern global
seafood marketplace. Unlike the Eastern oyster, which has
declined by 80 percent since colonization, the several species of wild Gulf shrimp are still with us in great numbers.
But the ramping up of the American shrimp appetite has
caused us to completely remake our seafood economy. Over
the course of the last fifty years Americans came to eat more
shrimp than any other seafood by a substantial margin.
Shrimp has become so popular that even the seemingly
boundless productivity of the vast marshes of the Mississippi River delta have come up short. It was this bottomless
American appetite for shrimp that caused us to start looking
beyond our shores for seafood and in turn compelled marine
scientists around the world to try to domesticate shrimp.
And it is because of this shrimp domestication project that
Americans today are able to cheaply consume more shrimp
per capita than the next two most popular American sea-

foods—salmon and tuna—combined. But as I learned when I traveled to Southeast Asia's churning shrimp-farming grounds, this transition from wild to farmed has come at considerable environmental and societal cost. In becoming Asia's premier market for shrimp, the United States has effectively unhitched itself from its own seafood supplies and hollowed out its ability and rationale to protect its own marine resources.

Lastly, the sockeye salmon of Alaska's Bristol Bay tell the story of the fights that lie ahead if we are to protect America's seafood future. The Bristol Bay sockeye salmon grounds are an untrammeled latticework of rivers, ponds, and lakes that can generate more than two hundred million pounds of fish per year. In spite of that seafood wealth, Alaska's leadership is seriously contemplating the permitting of Pebble Mine, a venture that would be the largest copper and gold mine in North America. If built, Pebble Mine would be the largest open pit mine in North America placed in dangerous proximity to the most valuable salmon fishery in the world. Saving Bristol Bay from this development requires that we recognize the importance of the sockeye salmon fishery that is endemic to the place. But while the fishery is critical to the fishermen who catch it, Americans outside of Alaska are only now starting to be aware that it even exists. The disturbing truth is that 79 percent of

Alaska salmon is exported, more and more frequently to Asia. Meanwhile, two-thirds of the salmon that Americans do eat is farmed and comes to us from abroad. This disconnect is a considerable obstacle to the fishermen and environmentalists working tirelessly to stop Pebble Mine: without valuing Bristol Bay as a food source, it's impossible to understand why it requires our defense.

But lest this book read simply as a litany of doom and destruction, it's important to point out that America, like a great ship that has been off course for so long, is ever so slowly coming about in the right direction. Forty years ago the Federal Water Pollution Control Act Amendments aka the Clean Water Act of 1972 sought to end the degradation of our estuaries, prosecute the polluters of our waterways, and protect this country's most vital seafood grounds from industrial development. Boldly it stated that it would "restore and maintain the chemical, physical, and biological integrity of the Nation's waters." From the Clean Water Act sprang a great flood of ideas and hopes that has affected every piece of seafood we eat and even the seafood we don't.

Through the act's requirement for cleaner water, the recovery of the Eastern oyster has become a stated goal of municipalities around the United States, even in estuaries as polluted as New York Harbor. A single oyster is capable of filtering up to fifty gallons of water per day—a fact that is not lost on state regulators. Indeed, more and more, oysters and clean water are being linked in the minds of policy makers, and this linkage could create a positive feedback loop that leads not only to cleaner water but to greater supplies of other seafood. It could even lead toward a sounder ecological approach to coastal protection in the face of sea level rise and ever more frequently occurring superstorms.

In the Gulf, the Clean Water Act has the potential to help repair a century's worth of damage the oil industry has visited on the shrimp heartland of the Mississippi delta. After spilling more than two hundred million gallons of oil into the shrimp-rearing grounds of Louisiana during the Deepwater Horizon disaster, BP could be liable for more than $20 billion in Clean Water Act fines and penalties. Properly applied, those funds could pay for the restoration of the Gulf's seafood heart and lungs: the extensive Louisiana marshes where 75 percent of the northern Gulf's seafood is born. Coupled with a growing seafood relocalization movement launched by bayou shrimpers, shrimp could end up being the motivation for saving the Gulf's marine ecosystems.

In Alaska's Bristol Bay, the Clean Water Act could gird those fantastically productive salmon grounds in a permanent defense against harmful industrial development. We have a chance to forever protect Bristol Bay, America's wild salmon epicenter, from becoming the site for a storage facility for billions of tons of mining tailings. Of all the fish fights that threaten our local seafood, nothing is bigger than the fight over Bristol Bay and Pebble Mine. Thanks to an unprecedented banding together of local native tribes, commercial fishermen, sport fishermen, conservationists, and chefs, the Pebble Mine fight might just be a fight that, for once, the fish will win.

But the future of the American catch depends not only on American governance, but also on the behavior of American consumers. There is no more intimate relationship we can have with our environment than to eat from it. Over the course of the last hundred years that intimacy has been lost, and with it our pathway to the most healthful of American foods. It is, in my opinion, our obligation to reclaim this intimacy and build a bridge from the plate back to the estuary. This requires us not just to eat local seafood. It requires the establishment of a working relationship with salt marshes, oyster beds, the natural flow of water from river to sea, and the integrity of the ocean floor. It means, in short, setting a new course that makes seafood not only central to personal health but critical to the larger health of the nation.

In heading out across the waters of my home city, my

home coasts, and my home country, and in telling about that journey, I'm hoping I might be able to convey how such a course might be set, to find the bases for our nation's broken relationship with its own ocean, and to understand finally how that breach might be mended.

Eastern Oysters

There are two kinds of oysters in New York City today: the kind that you can eat and the kind that you cannot. The first, edible kind are found in restaurants and raw bars, high-end grocery stores, and the few remaining markets that still specialize in fresh seafood. They are nearly always farmed and come to New Yorkers from a minimum of thirty miles away and usually from much farther, sometimes from clear across the country.

The second kind of oysters, the ones you cannot eat, are wild and live in New York City waters, within the thirty-mile blast zone of oyster destruction that now defines much of the metropolitan area's waters. These oysters, all of the single species *Crassostrea virginica*, are polluted, diseased, and staggeringly reduced from their former numbers.

And yet up until relatively recently there was no difference between the first kind of oyster and the second.

As late as the 1920s a local New York City oyster was the oyster that you ate. As Mark Kurlansky noted in his oyster-centric history of New York, *The Big Oyster*, up until the 1920s, the average New Yorker ate annually as many as six hundred local oysters as part of a locally sourced seafood diet of more than thirty-six pounds of fish and shellfish a year—more than double the current per capita level of American seafood consumption. New York oysters were so common as to be considered a poor man's food, priced at less than a penny apiece. If you happened to be wandering around lower Manhattan before the 1920s and were on a particularly tight budget, you could stop in at one of the many oyster saloons and purchase the "Canal Street Plan"—all the local oysters you could eat for a sixpence. And even though they were very cheap, cheapness alone does not explain their appeal. In the mid-1800s the average New Yorker spent more on oysters than on butcher meat.

In Search of Lost Oysters

On a recent July morning I donned a wetsuit, flippers, and an air tank and prepared to go scuba diving in New York City waters, in search of that second kind of New York oyster: the creature that was once the most local seafood of all, but that today is illegal to eat.

My dive was to take place at the very edge of the oyster blast zone in the last place in New York City that supported a local oyster industry. Known as Jamaica Bay, it was a tidal estuary straddling Brooklyn and Queens and bordered by the John F. Kennedy Airport. Around me, also wearing wetsuits, were a dozen inner-city high school kids from a newly formed institution called the Urban Assembly New York Harbor School. Also joining us were members of the city's leading marine nonprofit, NY/NJ Baykeeper, an organization whose mission is "to protect, preserve, and restore the ecological integrity and productivity of the Hudson-Raritan Estuary"—in other words, to make the wetlands and waters upon which New York City sits ecologically whole again. All of us had our own motivations for exposing ourselves to the hazards of scuba diving in a bay ringed by four sewage treatment plants and scores of industrial polluters. The Harbor School students and Baykeeper were there as part of a consortium of organizations that had just gotten approval to begin large-scale restoration of oysters to the New York side of the New York Bight. I was trying to find the root causes of the decline of local American seafood. And if we were to actually find a New York City oyster on this dive, it would inform our research, help us determine what was salvageable, and give us insight into what the future of the New York oyster might be.

Standing in the broiling heat on a litter-strewn beach on the edge of a now defunct airstrip called Floyd Bennett Field, we paired off with diving buddies. Mine would be César Gutierrez, a sixteen-year-old Bushwick native with yard-long curly hair pulled back in an explosive curly ponytail. Once we were suited up, the Harbor School dive instructor, a woman named Liv Dillon, handed me a compass, a measuring tape, and a knife.

"Do I have to take the knife?" I asked.

"Nah," said Liv. "But there's just a ton of crap down there, and if you get tangled up, César is going to have to cut you out."

I looked at César, who smiled and shrugged his shoulders.

Shuffling down the beach backward so as not to stumble over our flippers, César and I held on to each other by the wrist and entered the bath-warm water. To our left was the outflow of Mill Creek, one of the few remaining streams that still feeds directly into Jamaica Bay. Tidal river mouths are the favored place for oysters—the primary reason that greater New York once had among the biggest oyster beds in the world. Gotham is in fact one huge estuary. The city's many crenulated bays with variable water flow, the trickling of dozens of distributaries scattered over three hundred square miles, are what oysters love most. Once upon a time each little outflow had its own salinity and its own specific oyster taste. If land foodies rave about terroir, the qualities a particular place's soil gives to the tang of a local fruit or vegetable, oyster aficionados claim a *merroir*—the taste of home harbors carried in the watertight packaging of an oyster's twin shells. It used to be the case that you could actually taste the flavor of your New York neighborhood when you ate an oyster. Today most of the creeks of the Hudson-Raritan estuary have been interrupted, paved over, and rerouted underground, through the sewer system or trickling out along subway lines. Very little natural interplay occurs between rivers and tides today—one reason oysters have faded from New York waters.

Nevertheless, César and I were determined to find an oyster. Our mission was to measure out a twenty-meter line perpendicular to the shore and then work our way along the transect looking for oysters on either side. "Don't use your flippers too much down there," Liv had told us. "You'll just muck up the water. Go down there, establish neutral buoyancy, and just do this—" Here she crooked her index finger into a "C" shape. "That's all you'll need. Just pull yourself along the bottom slowly with your finger and look for the oysters."

Bobbing in the water now, we held the hoses of our buoyancy compensators above our heads and emptied our vests of
air. I felt more than a slight panic as my weight belt pulled
me down—something of the feeling Persephone might have
had when dragged by Hades into the underworld. All scuba
diving is a little claustrophobia inducing, even a dive in the
crystal clear Caribbean Sea, where visibility is usually at
least twenty or thirty feet. But here in New York City, as the
glow from the surface faded and we reached the bay's bottom, the world closed in around me in a dark and unsettling
way. In all, we could see only about eighteen inches ahead.
I was surprised when César began executing our appointed
tasks without hesitation. But these were his home waters,
the place he'd learned to scuba dive, and eighteen inches of
visibility was perfectly normal for him. He swam forward
and planted the anchoring stake of the transect. Then we
cruised on at an angle along the slope of the bottom, pulling
out tape as we advanced. At eight feet of depth, visibility
dropped to about a foot. I clenched César's wrist harder.
Down and down we went, ultimately reaching a point of
complete darkness.

Had large numbers of oysters been in place in Jamaica
Bay as they once had been, the quality of the water would
have likely been noticeably different. Oysters here at the
mouth of Mill Creek would have dealt with a lot of this

murk. River and creek mouths are so favored by oysters because the murk they carry is actually micro-algae—swarms of phytoplankton thriving on the nitrogen- and phosphorous-based wastes brought down from the land upstream. When these nutrients enter a bay, they fertilize the water and promote phytoplankton blooms. This phytoplankton is the oyster's principal food. Too much of it and the water will become choked and life will slip away. But when oysters are present, they gobble the phytoplankton down, cleaning and clearing the water in the process. A single oyster will filter as much as fifty gallons of water a day. Multiply that by several trillion—the historical oyster population of greater New York—and you can understand why the visibility in New York waters was once much better.

On we swam, past beer cans and tangles of fishing line and the occasional used condom. Had generations of our fellow New Yorkers been a little less abusive of our marine environment and instead allowed oysters to thrive, stands of eelgrass would have concealed all this trash. When a healthy oyster population filters and clears the water, sunlight is able to penetrate into the depths, in turn spurring the growth of several species of amphibious grass. This partnership of oyster and grasses turns out to be a foundational element of bountiful seafood in estuary environments. Oysters and marsh grasses stabilize the shoreline and create protective pockets of shallow water—essential for the sensitive lives of juvenile fish. This biological arrangement is known as a salt marsh, and its importance cannot be overstated. Salt marshes produce more basic food energy per acre than any other known ecosystem in the world—even more than tropical rainforests. They sequester more carbon than any other known ecosystem and can absorb as much as a foot of tidal storm surge. But where salt marshes really are worth their salt is in seafood production. Three-quarters of all the commercial fish species we eat rely on salt marshes for all or part of their life cycles. So it is in large part because of the oyster and the salt marshes they enabled that Dutch settlers arriving in New Amsterdam in the seventeenth century found vast schools of fish in New York waters.

And New York was not alone in this kind of biological configuration. Had César and I been swimming in the waters of any other Atlantic coastal city precursor before European settlement, we would have found the same abundant marsh systems. But nearly every time European colonists found a salt marsh, they would build a town, overharvest the oyster reefs, drain the wetlands, and fill in what remained. Even George Washington got in on the salt marsh draining game. In the late eighteenth century he formed a company to drain the mid-Atlantic salt marsh called by early colonists, rather indicatively, the Great Dismal Swamp. Today, as a result of all that draining, wild oyster reefs rank among the most endangered ecosystems in America, with 80 percent of the wild reefs now gone. Along with them have also gone the fish-producing salt marshes. By some estimates, the United States has lost 70 percent of its historical salt marsh, much of it in the last fifty years. Is it any wonder that the American catch is impaired by so much loss?

And so as César and I cruised through the water of Jamaica Bay, I saw nary a fish. Though Liv Dillon would later tell me that local divers like César, more accustomed to the water's low visibility, would probably have seen much more life, it was clear that Jamaica Bay was not the thriving fish nursery it once was. Remembering Liv's instructions, I flattened out my body and put a little air in my flotation vest. When I had achieved neutral buoyancy, I started pulling myself along the bottom with my left index finger. Slowly the water cleared enough for me to make out perhaps five inches of distance ahead. I groped along the bottom. It was mucky and cold and completely devoid of any recognizable structure. A flat, deserted plain.

Had there been oyster reefs of the kind that once paved more than four hundred thousand acres of New York's waters, the physical geography of the place would have been radically different and we would have scarcely been able to find a free patch of mud to plant our fingers. In addition to shoreside salt marshes, oysters that are farther out from land build reefs—a physical-biological infrastructure that is tremendously important for seafood. One of the reasons oysters are able to do this is their incredible fecundity. A single female oyster can produce up to one hundred million eggs. A salmon, by comparison, lays only about three thousand to ten thousand. Indeed, in the New York Bight of the past, in early summer, when oysters spawned, the waters probably at times would have had a milky cast, a living tide of embryos.

Only a few hours after being fertilized, oyster larvae become free-swimming "veligers" journeying in the water column until they detect, among other environmental signs, a chemically basic signal. When they get this signal they drop bottomward, find something good to attach to, and seal their fate for the rest of their lives. The clearest chemically basic signal for an oyster comes from old oystershell. And so once the first New York oysters established a layer on the seafloor, oyster begot oyster. Shell, or "cultch," as oystermen call it, is the foundation upon which natural oyster reefs form. Layer by layer the reef builds vertically, each new oyster generation building upon the last. No other bivalve builds in three dimensions with such architectural zeal. The closest thing to what an oyster builds is a coral reef—in fact, oyster reefs serve a very similar environmental function to coral. Just as tropical coral reefs create three-dimensional homes for swarms of brightly colored fish popular in home aquariums, oysters build reefs in temperate waters, creating homes for the kinds of fish we so treasure on our plates. According to some studies, the presence of an oyster reef can more than double the number of fish a given area of water can support.

Oyster reefs' fish-supporting capacity is complemented by the fact that sometimes oysters build such high reefs that they create protected slack water in their wakes—also ideal for fish nurseries. New York's Ellis and Liberty Islands are in fact pieces in a partly oyster-formed puzzle that stretched from Red Hook in Brooklyn a quarter mile out to sea across what is known as the Bay Ridge Flats. To some degree this was a natural seawall that was a first line of defense for Manhattan against storms as fierce or fiercer than 2012's Hurricane Sandy. If it is difficult to imagine oysters actually forming islands, consider the timescale: Oysters in a wild environment reach maturity at about three years, after which they spawn and generate young that will in turn form another layer of oysters. Repeat this process over the course of thousands of years, and the vertical accretion is considerable.

But beginning with the arrival of the first Dutch settlers in the 1600s, the oyster reefs of New York were mined out at progressively faster rates. And after the oysters were eaten, the shells were not returned to the sea. Primarily they were burned down and used as lime—a key component of mortar. Ironically, this mortar was at times used to help build sheer vertical seawalls at the ocean's edge—an environment that oysters find considerably more difficult to colonize than the gradual, crenulated slopes of the primeval Manhattan waterfront. In this way, oysters were used against themselves—melted down and recast as barriers to their own existence.

As César and I swam, we saw none of that old reef system. But the bottom was not entirely lifeless. I continued to drag myself along, one finger pull at a time. At a certain point, my finger started to hit something hard every time I advanced it. Pausing for a moment, I dug in excitedly, thinking that I had perhaps finally lighted upon a wild oyster. But it was another creature that I found—a hard-shell quahog clam six inches across. I inched myself farther again. Another huge clam. I inched again. Still another clam. The bottom of Jamaica Bay was literally paved with clams, all of them rendered entirely inedible by the last two centuries' pollution.

In spite of a growing sewage problem and all the damage wild oyster reefs suffered in the eighteenth and nineteenth centuries, oysters persisted as a local food source throughout the New York metropolitan area into the twentieth century mostly because humans decided to enter into a partnership with them. Overharvesting seriously reduced wild reefs by the 1820s, so from that time forward ambitious oyster entrepreneurs began importing juvenile "seed" oysters from still functioning wild beds in Long Island Sound, Delaware Bay, and Chesapeake Bay. By doing so they established a very profitable oyster-farming industry. They did this even though sewage increased exponentially through the nineteenth century into the twentieth. John Waldman, in his juicy, almost pungent biological biography of New York Harbor, *Heartbeats in the Muck*, noted that by 1910 six hundred million gallons a day of raw sewage were going into New York waterways. "Much of the bottom was covered with sludge," Waldman writes, "black in color from sulfide of iron; oxygenless; its fermentation generating carbonic acid and ammonia waste; and putrefying continuously, giving bubbles of methane gas." Furthermore, Waldman found that if you overlay on a map the areas where New York City historically dumped its raw sewage with the historical locations of its oyster beds, you would find that one had more or less replaced the other. The only surviving wild oyster beds in New York are those that escaped being buried under our muck.

Even so, the oyster industry in New York thrived. A study of New York City menus from the nineteenth century commissioned by the Wildlife Conservation Society found that the presence of oysters in New York restaurants rose continually from the 1860s through the early 1900s. By 1910 New York City produced 1.4 billion oysters a year. This despite the fact that New Yorkers were increasingly getting sick by eating them.

At the turn of the last century, scientists were only beginning to make the connection between water pollution, oysters, and illness. Dead Horse Bay, across from where we were now scuba diving, was so named because it was used as a burial grounds for old carriage horse carcasses after what could be salvaged had been turned into fertilizer and hoof gelatin. But that didn't stop anyone from farming oysters in that very same bay. In fact, as oyster grounds closer to Manhattan became more and more polluted, Jamaica Bay on the city's fringes grew to be one of New York's largest and most consistent sources of oysters until the 1920s. It was only when public health officials finally began tracing the origins of the era's major epidemics to oysters that the metropolitan bivalve started to lose its footing.

As the city grew and befouled itself, a primeval suspicion of the dangers of eating oysters came to be clarified by science. In 1854 the Italian scientist Filippo Pacini first discovered the bacillus bacteria that caused *Vibrio* cholera. In the 1890s a similar connection was made between bacteria and typhoid. Both of these bacterial diseases along with hepatitis A and B were found to be present in oysters in polluted waters. And as wave after wave of epidemics swept through New York City in the early twentieth century, oysters increasingly became identified as the vector for those diseases.

As the United States increasingly centralized regulatory power within the federal government, Washington started to take notice of what had formerly been a local health issue, taking particular note of oysters. In 1906 the federal Pure Food and Drug Act was passed, empowering a federal agency for the first time to restrict the interstate commerce of foods. Up until then, New York oysters had been a major export product. In fact, New York oysters were, in a way, the original exported packaged food. In the times before plastic, an oyster's naturally watertight shell served as a kind of biological shrink-wrap. They were shipped live across the country and even across the Atlantic to that other estuary city, London, which itself had also once had a lot of oysters but by the 1900s had very few. By 1896 New York was shipping a hundred thousand barrels of oysters a year to England.

But the seafood export life of the Gotham oyster was finally to come to an end in the twentieth century. A Chicago politician ate New York oysters and promptly died.

Eventually the New York oyster found itself under the Pure Food and Drug Act's jurisdiction. No more New York oysters were to be sold out of state. And soon after that, no more New York oysters were to be sold at all. Since no one could figure out how to make New York oysters safe or how to halt the flow of six hundred million gallons of raw sewage a day into New York waterways, health officials took the only action they knew would work. In 1921 they shut the Jamaica Bay oyster beds down. It was at that time that one of the first hints of replacing a domestic source of seafood with a foreign one was floated. "The stoppage of this supply of shellfish," New York's health commissioner Royal S. Copeland told the *New York Times* upon the beds' closure, "may mean that a supply must be imported either from Canada or from France to make up the deficiency."

With scant money and poor science to mitigate the effects of oyster pollution, it was easier to get rid of the oysters altogether than it was to combat the contaminants. And the mass closures of shellfish beds were to become commonplace throughout the coastal Atlantic. The same survey of New York menus that found a progressive growth of oyster dishes in restaurants from the 1860s through the 1910s shows a dramatic drop-off of oysters from the 1920s through the 1950s. By the 1960s very few Americans were eating any oysters at all. The industry followed suit. By the 1970s

American oyster production had plummeted to a mere 1 percent of its 1910 peak.

In places where local oysters had been officially abandoned as a food source, pollution continued unabated. The mobilization for World War II introduced a whole new range of industries to coastal zones that brought with them pollutants far more toxic than sewage. Heavy metals, phenols, pesticides—all of them were added to the trillions of gallons of sewage already entering American waterways. And the spread of industrial pollutants and the consequent decline in shellfish production occurred all over the United States. Long Island Sound, the Chesapeake Bay, and the oyster-growing grounds of Washington State all had epic industrial pollution problems, from the dumping of industrial dye by-product in the East Coast oyster fisheries to the spilling of mercury-laced pulp effluent from West Coast paper mills.

Yes, some mitigating developments occurred. In Jamaica Bay four wastewater treatment plants were constructed to try to stem the flow of effluent, but in a way they made things worse. Prior to the twentieth century much of New York's outer boroughs had no sewer systems—outhouses were the only means of getting rid of waste. As it turns out, outhouses are actually better for oysters than sewer systems. Outhouses cause bacteria to be filtered away by soil. Waste hits the sea only after it has been effectively neutralized. Sewer systems, meanwhile, conduct human waste directly into watersheds. The Jamaica Bay sewers were particularly bad. Instead of routing the effluent offshore into the open Atlantic, where pollutants would be more readily diluted, funding for the plants fell short and the outflows stayed in the bay, where they swirled about and deoxygenated the waters. A long-standing maxim of the wastewater control business is "The solution to pollution is dilution." In Jamaica Bay there was very little dilution.

Even today, despite the passage of the Clean Water Act in 1972, sewage remains a major problem. Though the law stipulates that all American sewage must receive primary and secondary treatment before entering a waterway, the truth is that New York, along with most American cities, never spends what it should on its sewers. Whenever it rains more than a quarter of an inch, New York City sewers are overwhelmed. In such instances the normally separate sewage outflow and storm water drainages merge. This same phenomenon happens in smaller towns and municipalities around the country, even municipalities with still functioning shellfish beds. The effect is comparable to a giant toilet bowl overflowing, with the waters of the nearest natural body of water becoming the receptacle of last resort. Every year these combined sewage overflows (CSOs in profes-

sional poop-speak) release hard-to-quantify amounts of un-
treated sewage with high levels of fecal coliform bacteria.
Today there are 730 CSOs throughout New York City, and
all hemorrhage when it rains. As far back as 1823 Thomas
Jefferson likened New York to the Roman goddess of the
sewer system, calling it a "Cloacina of all the depravities of
human nature." Poor Jefferson. He had no idea what was
coming down the pike.

César and I had some idea, though. Indeed, it was right
there in front of us, this terrible world of sewage and destruction. It pressed in around us as we plunged our crooked
fingers into the muck and pulled ourselves along, filthy clam
after filthy clam. But with as many clams as we found, we
could not find any oysters. Tracking back, on and on, brave
César and I searched. Not a one. Throughout the afternoon
we would scour the bottom, lifting rocks, pushing up along
the edge of the creek, diving deeper still off the edge of the
dredged channel into total blackness. Nothing. Nothing left
of a fishery that once hosted three trillion oysters, or of the
biological infrastructure they provided that played host to
the plentiful fish-holding habitat that kept average New
Yorkers nourished with thirty-six pounds of local seafood
a year.

Surfacing now, I thought of the immensity of the task
before the Harbor School, Baykeeper, and all the other

dozens of volunteers and organizations who had pledged to take on the restoration of the New York oyster. The multistate governing body New York–New Jersey Harbor and Estuary Program along with the Army Corps of Engineers had in 2008 declared the goal of restoring five thousand acres of oyster reef to New York Harbor in the next three and a half decades. How could this be done? How do you breathe life back into a dead food system? How do you connect the biological dots when so many of the dots have been impaired or destroyed entirely? Each and every element that supported oyster life in New York was gone. Each and every one of those elements would have to be rebuilt to restart the oyster anew.

These thoughts crowded my head as César and I began paddling over to the next dive site. But as we swam, clouds started to gather. A little while later a bolt of lightning split the hot, humid air. Then the skies opened up and a torrential rain came down upon us.

"Ahhhhhhhhhh!" the Harbor School kids screamed. The whole crowd of them tore out of the water, throwing their fins on the ground. Panting, they sprinted for the dive van, crammed inside, and slammed the door shut. Because of their education at the Harbor School, they knew what the average New Yorker does not: sudden, heavy rains meant the murky waters were only going to get worse. In a few minutes thousands of gallons of raw sewage would be flowing into the bay.

The Oyster's Argument
for Clean Water

Because we are a young country, we Americans carry within us an almost visceral sense of wild abundance used up and thrown away. It's for this reason that American nature enthusiasts believe in what I would call Environmental Original Sin—a sense that by being born into this world we have made this world dirty. It is a desire to purge ourselves of this sin that brought the New York oyster back into modern consciousness. If only we could reestablish the conditions that once made oysters thrive here, we could, like Dr. Seuss's regretful Once-ler, toss down a few Truffula-like oyster seeds and help the reefs regrow themselves.

But there is a tricky aspect to making the oyster the object of a 1970s-style environmental restoration movement. Because right now New York is in the middle of a twenty-first-century *food* movement, in which localness in cuisine is celebrated above all other characteristics. And even though most people who would like to bring the oyster back to New York City will grudgingly acknowledge that it is unlikely we will see edible New York City oysters anytime soon, and that it is the oyster's *ecological* benefits we need to restore to New York, not its culinary ones, there is a conflation of message and purpose that gets the oyster and the people who would restore the oyster into trouble. It's all well and good for New York City leaders to declare, as Mayor Michael Bloomberg did in 2007, the goal of planting a million trees in New York by 2017. No one is going to eat a tree, get sick, and die. Not so the oyster.

There is perhaps no person who understands this problematic conflation of environmental movement and food movement better than Andy Willner, the founder of NY/NJ Baykeeper and one of a handful of people who began the slow process of trying to bring oysters back to New York City. A little while after my scuba dive in Jamaica Bay, I met this bespectacled, white-bearded baby boomer in his hometown of Keyport, New Jersey, a town just a stone's throw from Lower Manhattan and at one time one of the metropolitan area's biggest oyster producers.

The son of aspirational Russian immigrants, Willner in his youth had seemed destined to follow in his father's footsteps and become some sort of respectable urban Jewish American professional. But after graduating from college in 1968, Willner unexpectedly dropped out of society and went to Alaska. "I drove across the Yukon and arrived at the town of Circle, where the road ended," Willner told me. "It was an eye-opener. I never saw that kind of wilderness before. Where you could go down to the river in the morning and be pretty much guaranteed that you'd catch a trout for breakfast."

When Willner returned to the New York area, he became a woodworker and sculptor and apprenticed himself to a boatbuilder. He began sailing around New York Harbor, poking around in its nooks, divining its nature. It was the 1970s, a time when people were just reawakening to the fact that Manhattan was an island and that the city was built on an estuary. The singer-activist Pete Seeger and his Hudson River sloop *Clearwater* had just started roving the harbor chasing down polluters. The Clean Water Act had recently been passed. All of this was in the air when one afternoon Willner found himself holding the idyllic memory of Alaska and the horrible reality of New York in his head at the same time.

"It was a really, really hot day," Willner told me. "I didn't have any child care and so I had to bring my daughter to the boatyard in Rosebank on Staten Island. She wanted to go swimming. She kept *begging* me to go swimming. And I started getting so angry that my daughter couldn't swim in the water. And I thought, Why the fuck *can't* she swim? If the Clean Water Act says that the water should be swimmable and fishable, why can't she swim?"

In fact, Willner was within his legal rights to press the government to answer such a question. Though the Clean Water Act had only just come into existence in 1972, it put forth that swimmability and fishability in American waters was to be the law of the land. U.S. waterway pollution legislation had begun much more timidly with something called the Refuse Act of 1899, an act that was directed mostly at the aesthetic horror Americans began to feel at the turn of the last century as they watched things like dead horses and rafts of excrement float past their swimming holes. It was a most idealistic law, generated by the same early twentieth-century Progressive Era mind-set that aimed to stop diseases like cholera and typhoid in urban neighborhoods. But like many idealistic laws, the Refuse Act lacked teeth and funding. It was unclear how polluters should be punished when they dumped unsavory things into waterways. It was similarly vague how municipalities would pay for wastewater treatment. In addition, there was the question as to which waterways should be covered under this first attempt at regulating water on a federal level. And while another attempt at national water quality control was made through the Federal Water Pollution Control Act in 1948 (along with six amendments made throughout the fifties and sixties), a nationwide federal-level consensus on the importance of clean water remained elusive. The environment being a mute and indefinite constituent meant that, more often than not, the health of our waterways was generally left to the states, which meant it was generally left to rot.

But by the 1960s, just as things like beaches and shellfish beds were reaching their ecological nadirs and people started saying the word "ecology" out loud, attitudes toward both the environment and the role of the federal government began to change. The large wartime-era industries began to go a little too far in the size and scope of the polluting projects they proposed. Rivers caught on fire. Fish kills from deoxygenated waters abounded. In 1968 a nationwide testing of 590 water samples by the U.S. Fish and Wildlife Service found that 584 of them contained the carcinogenic pesticide DDT at levels exceeding the limit set by the Food and Drug Administration (FDA), sometimes as much as nine times greater than the limit. In 1971 the FDA found that 87 percent of swordfish sampled had mercury levels well above government-mandated maximums. Alongside these developments, a U.S. Supreme Court that took environmental goals seriously came to power. In the landmark 1966 case *United States v. Standard Oil*, the court ruled that the 1899 Refuse Act could be interpreted to mean that industries were to be prohibited from discharging *any* substance into navigable waters unless they had first obtained a permit from the federal Army Corps of Engineers. In other words, you could not pollute American waters unless you had the federal government's permission.

This decision would greatly inform legislators who took up the drafting of the Refuse Law's offspring: the Clean Water Act. When that legislation was finally signed into law in 1972—after Congress overrode President Nixon's veto—it made a staggering promise: "To restore and maintain the chemical, physical, and biological integrity of the Nation's waters." In order to achieve this objective, the law began with a preamble that is so bold in contrast to today's milquetoast policy making that it is worth repeating again and again: "it is hereby declared," the law stated, "that the discharge of pollutants into navigable waters be eliminated by 1985." And the act had financial teeth. Not only did it have the power to fine polluters for their transgressions and to use those fines to further the goal of clean water, it also included a federal grant-making program that mandated that billions of dollars be invested into building and updating sewage systems. Very quickly American waterways started getting cleaner. Fecal coliform levels in New York decreased by a factor of ten. Levels of dioxins in once commercial seafood species like striped bass and blue claw crabs dropped by half in just ten years. Water oxygen levels rose to be higher than they had been in a century—double what they had been during New York's dark days in the early 1900s. Levels are now regularly above the 3.2 milligrams per liter needed for oysters to survive and often above the 5.5 milligrams per liter amount required for oysters to thrive. In all, the Clean Water Act achieved "swimmable and fishable" conditions in 65 percent of the nation's waterways.

But 65 percent wasn't good enough for Andy Willner. Unlike the political climate of today, where legislative promises are viewed with deep cynicism if not apathy, people involved with social movements in the sixties and seventies—people like Willner—took such promises seriously. Willner knew from his time in Alaska what swimmable and fishable water looked and smelled like. And he believed you could and should have the right to safely eat from all American waters—even New York City waters. Soon after his experience with his daughter on Staten Island, he founded the organization that came to be NY/NJ Baykeeper. He started to track down industrial polluters all over the New York Bight. He wrangled journalists and marooned them on boats for hours until they felt as viscerally sick about the water as he did. He filed lawsuits right and left. He got so good at tromping around the city's remaining wetlands and detecting pollution by the smell in the air that at one point a local judge designated his nose an expert witness.

But at a certain point he realized that the scope of violations was so numerous that a single individual or even a single organization could not possibly track and prosecute them all. And this corresponded with a time when he happened to be patrolling in the Baykeeper boat on the Navesink River, a tributary to Raritan Bay in New Jersey, with a young activist named Ben Longstreth. Rounding a bend in the river approaching Red Bank, they glanced at their nautical chart and saw they were at a spit of land called Oyster Point. "I wonder what it would take," Willner recalls Longstreth asking aloud, "to bring oysters back to Oyster Point?" This set off some serious light bulbs. Why weren't there oysters at Oyster Point? Where had they gone? Why was the only remnant of them a note on a nautical chart? Obviously, if there were oysters there, oyster lovers from New York and New Jersey wouldn't be able to eat them. But the Clean Water Act stipulates that all American waterways should be made fishable and swimmable by 1985. Willner recognized that the law clearly stated that a New Yorker or New Jerseyan had the legal standing to demand that they be edible. Or as Willner puts it: "If you can't eat the oysters, why the fuck can't you eat the oysters?"

This question bothered him intensely. By the time Baykeeper was founded, most New Yorkers had completely forgotten about their city's oyster past. How could they be reminded that they once had oysters everywhere? And how could an oyster culture be rebuilt to remind them of this point? Curiously, the means to an answer to this question lay in oil.

"Throughout the 1990s there were eight major oil spills in New York Harbor," Willner told me. "New York became an expert in oil spills. We became experts in how to engage." A major part of that engagement, Willner decided, was to use the Clean Water Act damages obtained from polluters to fund oyster restoration. It was a solution fitting to the spirit of Willner's generation of activists: to use the power of the forces of bad water and redirect them toward helping the forces of good water.

As a first attempt, Willner decided to see whether he could establish some wild oyster beds simply by re-creating the conditions the harbor used to have. After a 1996 heating-oil spill in the Arthur Kill, Baykeeper supplemented seed funding from the National Fish and Wildlife Foundation with oil spill damage money and began laying down oystershell in fifty different locations throughout the New Jersey half of New York Harbor. Since Willner knew a bit about oyster biology, he had speculated that if you put down some shell on the seafloor, the remnant oyster larvae still in the water might somehow find those little patches of shell, set, and perpetuate.

It didn't work. The patches of shell were too few and far between. The number of oyster larvae in the water was infinitesimally small compared with what it had been. The chances of viable New York oyster larvae finding introduced shell was the marine equivalent of trying to land a space probe on Mars without the benefit of Mission Control.

But this came to be understood only in retrospect. "At the time," Willner said, "we thought, well, maybe the water's not clean enough for oysters yet. Maybe we have to see if they'll survive. So we decided to see how well oysters would do if we just put them out." Launching a program of "oyster gardening," Willner got several dozen volunteers to put out two-foot-by-three-foot mesh cages full of juvenile oysters all over the New York Bight. Oyster gardening is a phenomenon that had been steadily cropping up throughout the country in the wake of the Clean Water Act, especially in areas where the water became clean enough for oyster gardeners to eat the oysters that they grew. In Washington State's Olympic Peninsula, for example, Taylor Shellfish Farms, the West Coast's biggest shellfish producer, rings its oyster farms with armies of citizen gardeners as a first line of defense against municipal water pollution. "When they are buying their shellfish seed at our seed sales we tell our gardeners," Taylor's

Bill Dewey told me, "if they want to eat their oysters, they may want to check and make sure their septic system is working, and they may want to consider picking up after their dog, and think about what types of pesticides they are using on their lawn." An added benefit to Washington's shellfish gardening efforts is that Taylor Shellfish Farms enlists volunteers from local water-quality nonprofits to help at the seed sales and donates a significant percentage of the sales back to them for projects in the watershed.

But in New York the gardeners weren't planning on eating their oysters and there was no oyster industry with which to collaborate. And so the gardeners made themselves a very simple goal: to merely see whether the oysters would even survive. From the Battery to Harlem, across the Bronx to City Island and Soundview, east to Jamaica Bay, south to Red Hook, and back west to the Arthur Kill and the Kill Van Kull and on into the bedroom communities of Newark, these cages, tiny sentinels of the lost age of the oyster, were put out in the wastelands of their former kingdom. "We put out oysters in fifty different places," Willner told me. "And it was astounding how well they did. They survived everywhere. Well, not everywhere. They died in the Gowanus Canal. But they didn't die from the pollution. They died because the rats ate them."

Once he knew that oysters could survive, Willner engaged a group of scientists to see what would really need to happen to get oysters to self-perpetuate, to form the reefs of yesteryear and take over the difficult work of rebuilding themselves and cleaning the water. "The scientists said you've got to really have a huge amount of shell in the water for this to work. Just a gigantic amount of shell. And so we decided to try that." In perhaps the New York oyster's best photo op in modern history, Willner used funds obtained from oil spill settlement payments to pay for the dumping of a spectacular pile of shell at Liberty Island right beneath the legendary statue in 1998. "We brought a barge up from Chesapeake Bay," he told me. "We had hundreds of people. It was a landmark event."

But the Liberty Island project didn't work either. There was too much sedimentation and too much current in the modern-day harbor. The nascent reef drifted away into the open Atlantic.

Finally, the scientists advising Willner told him to take the project up another level. "The scientists said you had to *build* reefs," Willner explained. Looking through a few different possibilities, the oyster restorers decided on a design that had recently been developed to rebuild coral reefs. Such structures used hundred-pound blocks of concrete studded with juvenile coral polyps as anchors for what was hoped would grow into much larger, self-growing structures. Willner was advised that if he used juvenile oysters instead of coral polyps—something called "spat on shell"—oysters just might take. He soon brought in Pete Malinowski, the son of a Long Island Sound oyster grower, to oversee the project. Malinowski had recently joined the Harbor School faculty and was trying to figure out some practical way to direct his students. On the Hudson River in Manhattan's Chelsea neighborhood Malinowski constructed a basic, hand-built version of what is known in oyster-growing jargon as a "flupsy"—a floating upweller system—basically an oyster nursery that grows oysters out to plantable size. For Willner and Keyport Harbor, Malinowski designed "reef balls"— concrete masses that oysters could adhere to and grow on. After a year of grow-out, the reef balls were placed at the bottom of Keyport Harbor, within sight of the New York City skyline. It was just a quarter acre in all.

Boy Scouts came to plant the one-hundred-pound reef balls, and by 2010 oysters finally began to take hold at the bottom of Keyport Harbor. The network of private citizen oyster gardeners that Willner had helped to establish a few years earlier began motoring out to the Keyport reef with their children and hand dropped the oysters they'd reared in their oyster garden cages down onto the reef.

"Some of the kids had separation anxiety," Willner recalled. "I mean, they're like rocks, really, but the kids were attached."

But all that was to end abruptly.

On a Friday afternoon in the summer of 2010 a fax arrived at Baykeeper's office from the New Jersey Department of Environmental Protection stating that Baykeeper had violated the terms of its oyster restoration permit. In the previous year Governor Chris Christie, a probusiness Republican who had wanted to make a clear break with the previous Democratic administration, had come to power and replaced oyster-restoration-friendly officials with fresh appointees. The new administration appears to have been troubled by fears raised by oyster growers elsewhere in the state. Southern New Jersey is among the United States' largest commercial shellfish producers, and health scares around seafood are remarkably transitive. So seared into the American subconscious are the pollution-caused seafood plagues of the early twentieth century that fear easily slips irrationally from one piece of seafood to another. As one oyster grower put it to me, "A rotten lobster getting someone sick in Maine can make Americans in Florida stop eating pompano."

Even though Baykeeper had expressly noted that the Keyport oysters were not intended for human consumption, New Jersey was not convinced. The oysters on the Keyport reef, the legal brief stated, were in fact an "attractive nuisance" as defined by an obscure statute. They were being grown in polluted waters within range of possible human harvest and thus a hazard to public health. In a covert raid on the reef, Department of Environmental Protection officials had discovered that a few of the oysters had grown to a market size of three inches and could conceivably end up in the New Jersey food system if someone were to poach them. Oyster restoration and research, even oyster gardening on the New Jersey side of the New York Bight, the state authorities concluded, had to stop *immediately*. Baykeeper was instructed to remove each and every oyster from the bottom of Keyport Harbor within ten days or face a fine of ten thousand dollars per day—a death knell for a tiny citizen-supported nonprofit organization with a total annual budget of less than half a million dollars.

Though Willner had already retired and passed the Baykeeper torch on to a former New Jersey regulator named Debbie Mans when the action took place, the aggression of it all seared him. "We were hit with the sledgehammer of the state," Willner recalled with 1960s-style hell-no-we-won't-go anger. "It was a cheap shot, sending that fax on a Friday afternoon. Where the fuck are you going to find a judge on a Friday?" Debbie Mans and the oyster restoration program director, Meredith Comi, had no other choice.

After scrambling to find funds, the organization hired a barge to remove the oysters from the bottom of the harbor. Onshore the Baykeeper staff chipped the oysters off the reef balls and left them out to dry and die. Then the remnants of the two-hundred-thousand-dollar project, the broken shells and twisted meats, were taken across a parking lot and thrown into a Dumpster.

Stepping out of Willner's car now at Keyport, we walked over to the dock where the destroyed reef balls now sat. "This is what the oysters were attached to," Willner said, mockingly trying to heft the weighty chunk of concrete. "Can you imagine diving down and trying to poach from this? The oysters were all growing inside. You couldn't even reach them underwater. That was what was so frustrating. They just didn't get it. This wasn't going to be a harvestable reef. It would have been a *real* reef."

Taming the Oyster

With New Jersey's shutdown of the Baykeeper oyster program on the western side of New York Harbor in 2010, the center of gravity for oyster restoration has shifted over to the New York side of the bight, with a focal point being the

Urban Assembly New York Harbor School, the home base for the students who took me for my oysterless oyster dive in Jamaica Bay.

The Harbor School is an unusual institution that makes its home on the tiny clump of land between Manhattan and Brooklyn called Governors Island. There, on this little speck of land that Russell Shorto in his book *The Island at the Center of the World* identifies as the site of the very first Dutch settlement, the oyster restorers have funded and installed what is probably the most crucial missing piece for the New York oyster's comeback: a means of oyster reproduction. With nearly all of the New York salt marshes gone and with most of the reefs plundered, an outside biological jumpstart is needed to get larvae into the water again. Pete Malinowski, the school's oyster program director, had built a provisional nursery two years earlier, but the one he created at the Harbor School was the real deal—a professional operation with big plans. The stated goal for the nursery is to generate one billion oyster juveniles that would be planted in the New York Bight. It's a seemingly impressive number—that is, until you hold it up against the natural oyster nursery that was once the entirety of the New York City estuary. In colonial times, when three trillion oysters populated New York City waters, the annual spawn of oyster larvae probably numbered somewhere in the environs of 300,000,000,000,000,000,000,000, or, more readably stated, three hundred quintillion.

Nevertheless, Malinowski, a tall, blond thirty-year-old with an iron grip and a sea blue gaze, is plowing ahead in the only way he knows how. Malinowski hails from Fishers Island, a sliver of land off Connecticut a hundred miles east of New York City. While other boys in this affluent Long Island Sound enclave spent their summers in the 1990s chasing girls and driving BMWs, Malinowski and his four siblings worked long hours at their father's oyster farm, chasing oyster larvae across microscope slides and building up an oyster production business that is today well known up and down the Connecticut coast.

In fact, had it not been for the Malinowskis of Fishers Island and the handful of Connecticut oyster folk who persisted in the wake of the twentieth-century oyster apocalypse, the oyster industry in the Northeast might have collapsed entirely. Connecticut, I am proud to say as one of the state's native sons, is something of a refuge of last resort for the New York oyster. Nearly a century of Connecticut-based oyster science is a major reason that today the Northeastern oyster industry outside of New York is starting to stage a comeback. As recently as the 1960s, commercial oyster production in the United States had been pummeled down to a fraction of its former strength. Though the East Coast oyster industry remains tiny in comparison to its heyday at the turn of the last century, production has doubled in the last twenty years. All those dollar-apiece oyster specials New Yorkers are finding at happy hour of late are a result of the industry's resurgence and in large part a legacy of the advances made in Connecticut. And why Connecticut in particular can claim pride of place in the oyster's modern resurgence has to do with the migration of a peculiar individual to the state in the 1920s: a tsarist-era Russian who, with a tsar's dictatorial power, helped to tame the oyster.

The great oyster tamer's name was Victor Loosanoff. He grew up in Moscow, raised in a Russian aristocratic culture that favored frequent pairings of Champagne and oysters. But Loosanoff's lifestyle was cut short in 1918 when Lenin seized power. After briefly fighting the Bolsheviks, he fled across Siberia, made his way to Alaska with help from the Russian community in Harbin, China, and then paid his way south to Seattle by working on fishing boats and bare-knuckle-boxing through the savage lumber camps of the Pacific Northwest. After teaching himself English he alighted at the fisheries program at the University of Washington. It was there that he fell in love with shellfish in general and oysters in particular, a passion that would lead him to the U.S. government's emerging fisheries development programs and, in 1931, to a ramshackle research outpost in Milford, Connecticut.

New Haven and, by association, Milford had up until the 1920s been at the center of an incredibly profitable oyster industry. Sail-powered oyster sloops regularly queued up around the large wild oyster beds at New Haven and Bridgeport and transported the bounty to the New York markets. The trade only increased as the New York beds started to fail. But by the time Loosanoff arrived, all that was fading. Connecticut oystering was spiraling downward. Pollution, fear of disease, and all the other factors that had afflicted New York were now spreading into a much larger oyster blast zone.

Loosanoff, once installed in Milford, set about building a shellfish research operation and true to his early military discipline, he ruled with an iron fist. He developed dozens of different lines of inquiry and assembled a staff that cowered at his approach. "He would make the rounds every day," a chief of research at the Milford lab told me. "If you didn't have something new when he came and visited with you, he would throw a fit. He was a very hot-tempered individual." Upon his retirement, one bullied researcher assembled all of Loosanoff's published scientific papers in a single gilt-edged volume embossed with the private nickname employees had given the author out of both respect and scorn: "Le Bâtard."

But Loosanoff's brutal demeanor yielded results. In particular, what Loosanoff significantly contributed to was the methodology of growing oysters in a nursery outside of ruined nature. As wild seedbeds became more polluted and less capable of supporting the industry, Loosanoff realized that some way had to be developed to take fragile oyster larvae out of the wild and grow them reliably into plantable seed. It was under Loosanoff's direction that researchers combed through hundreds of species of algae and determined which of those species were best for nourishing young oysters. Loosanoff helped the researchers figure out how to isolate those algae species and grow them in a sterile culture that could then be used as starter food for oyster nurseries everywhere. This technique, which came to be known as the "Milford Method," was to become a key engine for oyster aquaculture throughout the world.

Loosanoff's techniques, invented in the thirties and forties and refined in the fifties and sixties, turned out to be something of a magical catalyst when combined with the improving national water quality in the seventies and eighties brought about by the Clean Water Act. Cautiously, old oyster growers started upping their production. New growers from Virginia up through Rhode Island, Massachusetts, and Maine put a toe in the water and were struck not only by the decent money they could make off oysters but also the positive change they were bringing to the marine environment by farming them. "We took a patch of 2.3 acres of black, eutrophic, anoxic bottom—unproductive and virtually barren," Rhode Island oyster grower Bob Rheault wrote me recently about Moonstone Oysters, a farm he founded near Point Judith in 1986. "And after a few years it was transformed into one of the most vibrant dive spots in Rhode Island. We sent a few scientists to document the diversity and abundance of species. We found that our pissant little 2.3-acre farm was now home to a thousand baby lobsters, thousands of juvenile sea bass, and other fish. We set up traps for scup (a New England commercial fish resembling a Mediterranean dorade) around the perimeter because fishing on the lease was so good. We slammed them. We would catch keeper striped bass on the lease many months of the year, attracted to the lease by the forage fish and structured habitat around the oyster cages. When I took my kids fishing we went to the lease because I knew we would catch fish."

But in spite of the incredible ecological restorative powers of oysters and oyster culture, oyster growers all over the United States and indeed anyone trying to grow anything in the water are facing an ever more difficult uphill climb. One reason for this is that the places that are good for growing seafood tend to be the places where the wealthiest Americans would like to build their country houses. Back before the 1920s, when the shellfish industry still wielded significant economic power, oystermen controlled large portions of both the East and West Coast shoreline. In fact, some of the first cases of U.S. property law stem from disputes between oyster growers on the Connecticut coast. But once disease and pollution pulled the economic rug out from under the growers, the ability to own and control valuable waterfront property began to slip from the industry's fingers. Now, all over the United States, sites that would be ideal oyster locations are simply unaffordable. Moreover, wealthy shorefront landowners are becoming ever more aggressive in actively keeping oyster growers from returning.

"The opposition we have run into related to getting new farms permitted is largely related to impacts to people's viewshed," Bill Dewey of Washington's Taylor Shellfish Farms told me during a quick trip I made to the West Coast. Beginning in the 1990s, when Internet millionaires began abandoning Seattle for the greener, quieter environs of the Olympic Peninsula, they began buying up the rugged shorelines and installing sweeping seaside estates. As Dewey put it, "These newcomers sometimes don't appreciate the value of a working waterfront and the ecological services being provided by shellfish on the farms. Instead they value pristine views and unobstructed waters for their boats and jet skis. An added challenge in Washington State is that winter low tides occur at night. If someone buys a home in the winter, they might not realize there is a commercial shellfish farm adjacent." Oblivious to the oyster leases that surrounded their new property, the tech barons would build their fancy houses all winter only to find themselves surrounded by shellfish growers when the daylight low tides arrived in spring.

In a weird perversion of environmental legislation, landowners trying to clear their shores of all those shellfish farmers tried unsuccessfully to sue growers like Taylor Farms under the Clean Water Act, claiming that their shellfish farms were introducing shellfish feces into the marine environment. Fortunately the courts recognized the fact that shellfish farms tend to have the effect of a net *reduction* in nitrogen and other pollution. This user conflict phenomenon represses oyster abundance and the oyster business all across the country. As a Connecticut oyster grower named J. P. Vellotti put it to me, "It's nearly impossible to start an oyster operation here when there are million-dollar houses all around me."

And with unfriendly shore owners mucking around in the permitting process, even more beneficial American seafood industries have been senselessly held back. Bren Smith, a grower on the Connecticut coast near the town of Branford, told me how getting a permit to farm kelp proved to be one of the most difficult hurdles of his life. Smith had turned to kelp after having his crops of oysters repeatedly blighted, either by hurricanes that buried his oysters under mud and muck or by the frequent shellfish closures that occur because of wastewater treatment failures nearby. Kelp turns out to be an extremely efficient and easy crop to cultivate. It grows three feet a month, removes more nitrogen waste from the water than oysters (or pretty much anything else, for that matter), is higher in protein than some fish, and rich in omega-3s. But in spite of these incredible benefits, getting a permit to farm kelp required Smith to complete two hundred pages of paperwork and wait two years. This could be one reason that 98 percent of the $5.7 billion world seaweed market is concentrated outside the United States. Mussels, another great water cleaner and omega-3-rich seafood, are similarly difficult to permit. It is no surprise that they are another U.S. seafood deficit item. Of the $108 million in mussels Americans eat, 90 percent are imported.

Nevertheless, American coastal seafood growers have persisted and in some cases even thrived. Some growers have succeeded in persuading wealthy landowners of the clean-water benefits of shellfish. Steve and Sarah Malinowsky, the parents of the Harbor School's oyster program director, Pete Malinowsky, have been particularly effective in this regard. On their oyster farm on Fishers Island, they managed to persuade their neighbor, an heiress to the DuPont estate, to allow their oyster cages to impinge upon her viewshed. "Oysters add distinction to my property," she is reported to now say proudly.

And so the committed few soldier on and spread the oyster fever with a missionary zeal. "I just love the idea of populating the East Coast with oysters," Steve Malinowski told me enthusiastically when I asked him about his family's work. And his love, it seems, is expressed in the fertility of his accomplishments. Not only has he fathered five children (two of whom now work at the Harbor School); he has also established an oyster engine that churns out something like thirty to forty million seed oysters a year. Indeed, many of the oysters being introduced to New York City are of Fishers Island stock, and a good portion of the resurgent growers that are slowly reviving the oyster industry up and down the New England coast also owe a debt to the Malinowskis.

And so, beginning with Loosanoff's Milford lab, a line of oyster pedagogy can be seen developing over the years, passing through Steve Malinowski to his son Pete and finally to the Harbor School kids, whose course of study, it is hoped, will ultimately prepare them to be new guardians of the city's oyster culture. A culture that will, in time, help fuel the repopulation of New York waters with fish.

Visiting Malinowski's aquaculture class at the Harbor School, I was able to see some of the adult Fishers Island oysters sitting in their tanks getting ready to spawn. There, the sixteen-year-olds titrated the microscopic larvae, studied their health, and filtered them through screens to grade their size, mixing and matching, trying to figure out which of all these young oysters might be the right ones for the New York Harbor of the twenty-first century. And there would be ample opportunity to try them out. While New Jersey had prohibited additional oyster gardening and artificial reefs, New York had decided otherwise, with its plan for putting in place five thousand acres of oyster reef by 2050.

The only problem is that everything the Harbor School is doing—the breeding of oysters and planting them onto test sites throughout the New York Bight; the careful monitoring, reseeding, and expansion—is proceeding very slowly. The introduced oysters are dying left and right. How and why this is happening has to do with the oyster's next major obstacle: the physical environment of the waterfront.

Which is the very thing that is making the city vulnerable to superstorms.

Oyster-Tecture

"There isn't going to be some kind of Kumbaya moment when nature returns to the world here in New York and humans sway along and embrace it." So said Kate Orff, a Manhattan-based waterfront landscape architect who has been one of the New York oyster's most visible advocates for the last five years. A student of the world-renowned architect Rem Koolhaas and a longtime collaborator with the Harbor School, Orff has a Harvard grad's confidence and a downtown denizen's sophisticated bearing. A little while after visiting the Harbor School oyster nursery, I dropped by her landscape architecture firm, SCAPE, to talk about oysters. But not to talk about them as food or as biological organisms. Rather it was Orff's idea of oysters as architecture that intrigued me. The idea that oysters could bring New York City to a different aesthetic place that was not only closer to the marine-centered ways of the past but also responsive to the rising sea levels of the future.

Orff is the coiner of the term "oyster-tecture," and on the invitation of New York's Museum of Modern Art in 2009 produced a schema of a reimagined New York City seascape in which fully restored oysters would play a critical role in making New York City's shores greener and more protected. Over the course of the next fifty years, the sea level around the island of Manhattan is predicted to rise approximately two feet. Battery Park, the far West Village, the South Street Seaport are marked for inundation. About 20 percent of New York City is built on landfill—landfill that was dumped on top of oyster grounds. All of that landfill and ruined oyster country will eventually be reclaimed by the sea. If the predictions prove accurate, nothing—not sea-walls, not pump stations, not anything—will stop that.

But Orff believes that the oyster is reminding us of our watery past while telling us something about our underwater future. What she thinks New York should become is a wetter place, more tidal and more reminiscent of the shifting water-soaked environment it used to be when it was more estuary than city.

In such a scenario oysters wouldn't stop the inundation of New York. But they could serve to mitigate its extremes. If planted in a bay, oysters could over time accrue and cause it to become increasingly shallow. In Jamaica Bay this could have important storm-mitigating effects. The energy in tidal storm surges tends to dissipate when currents pass over shallower water and is further dissipated by the dense assemblage, or "roughness factor," of a reef. It's just this kind of shallower, calmer water that oyster beds could enable and propagate. Other buffering possibilities exist. An oyster reef could, for example, extend the breakwater at Breezy Point at the far end of the Rockaways and shield Coney Island in its lee. Other oyster-tecture projects are even more ambitious. In one scenario Orff proposed rebuilding the vanished oyster reef that once stretched across the Bay Ridge Flats from Red Hook to Governors Island. Once in place, the reef would rise higher and higher, eventually providing a buffer against open swells for parts of Brooklyn and Lower Manhattan.

Orff has also imagined a way for these reefs to be self-perpetuating: a biological reinvigoration of the Gowanus Canal and Newtown Creek.

Today both the Gowanus Canal and Newtown Creek are listed as Superfund cleanup sites (bodies of water that are so polluted that the marine scientist John Waldman calls them "punched senseless by man"). But their geographical position and existing urban infrastructure opens up an interesting opportunity. In 1911 New York City built a flushing system for the Gowanus that had the possibility of moving a lot of the already bad water out into the open sea and reoxygenating the canal in the process. During the city's decline in the 1960s, that flushing system broke and remained out of commission until the New York Department of Environmental Protection reactivated it in the late 1990s. It again proved inadequate and was shut down in 2010 for upgrading. It is expected to restart again in 2014. If the machinery can truly be restarted Orff imagines that a series of oyster nurseries could be placed throughout Gowanus and Newtown, next to the flushing system, allowing oyster larvae to be blown out toward the Bay Ridge Flats. There they would meet up with larvae coming from the other direction, from Pete Malinowski's nursery on Governors Island, and if the right structures were put in place in the middle they would converge to form an über-reef.

But what kind of structure would that be? It turns out that in order for oyster-tecture to work, the damaged and changed geography of the harbor itself will have to change.

The natural depth of most of the waters of the New York Bight was at the time of colonization about twenty feet and the water flow at the various creek mouths was slow and flaccid. Large parts of the bight were shallow, shellfish-friendly flats, often exposed to the open air at low tide. But eventually Americans outgrew their oyster-rich shallow harbors. Between 1830 and 1855 the tonnage of the average vessel entering New York Harbor quadrupled and the amount of water those new vessels drew increased by 25 percent. Many of the bays and inlets of New York were made progressively deeper to accommodate them. Today large portions of the bight have been dredged to fifty feet and the increased depth has caused currents to flow far more swiftly and sediment to pour down like never before.

"What we learned when we started research on this was that because of the dredging of the harbor and the increased current speed, oysters that settled on the floor would be smothered by silt and die," Orff explained to me. "So what they need is a new bathymetry, a false floor to lift them up over the sedimentation zone. That's where oyster-tecture would work—it would be a way of coming up with an armature that would be literally something that oysters can hang on to. We would basically build starter reefs from piles of interlocking quarry rock bolstered by knitted fuzzy rope to create a foothold so they could begin to grow and change as they grew." In oyster restoration circles one hears a lot about "fuzzy rope," a shellfish aquaculture technique that uses the surface area of a frayed rope to attract and encourage oyster and mussel larvae to set. In Orff's imagination, a mosaic network of shallow bathymetry would begin to cover the bottom of New York Harbor, which would support and encourage oysters to grow. "We've begun to work with the Stevens Institute's hydrodynamic water circulation model to see where these breakwaters and reefs could be effectively sited, and how best to build them," Orff told me. "Compared to other approaches, it's relatively cheap. We could literally have a bake sale and we could get started."

Beyond rock and fuzzy rope, oyster-tecture might also encourage planners to rethink the city's most basic element: concrete. It turns out that concrete itself can be reformulated, coated with calcium carbonate, and given an alkalinity that would draw in oyster larvae to set. Oyster propagation could be even further aided by the city's restaurants. The average New York City dining establishment discards tons of shell a year. All of it is thrown into landfills. All of it conceivably could be put back into the water to make cultch— the chemically basic thing that oysters like most.

It all sounds good. But it also made me think that the oyster can serve many agendas. Baykeeper's Andy Willner had used the possibility of an edible New York oyster as a way of pushing forward the enforcement of the Clean Water Act even though he knew that New York oysters would be next to impossible to make edible. Here it seemed as if Orff was employing the dream of oysters as coastal infrastructure as a way to push forward the idea of a "softer," biologically oriented shoreline. But as utopian as it might sound, Orff feels there is very little alternative. "It seems to me we have to collectively try to reimagine cities and natural systems coexisting," she told me. "If you can't imagine that, I'm not sure what kind of world we are going to be living in fifty years from now."

What Orff says rings right, especially when you look at some of the truly grotesque abiological plans some engineers have for the city. In the winter of 2012, six months before Hurricane Sandy hit New York, the Army Corps of Engineers was contemplating the construction of an enormous seawall that would extend five miles from Breezy Point across the harbor to Sandy Hook, sealing in New York like a cork. It would take many years and many billions of dollars to build. The construction of oyster reefs and the "flupsification" of the Gowanus, by comparison, would cost under $1 billion.

And at the end of all this, at the conclusion of a major oyster-tecture effort, would New Yorkers ever again have an edible local oyster? Amid the bathymetric readings and fuzzy rope schemata that cover SCAPE's walls is a timeline showing the different oyster-tectural milestones that could be accomplished over the course of the next fifty years. At the end of the timeline, sometime around 2050, a woman in sunglasses stands at the end of an oyster-tecture pier in Red Hook. She cups a Gowanus-spawned oyster in her hand, ready for a healthy slurp.

Gotham *Merroir*

Oyster-tecture, the kind of vision for the harbor that Orff rendered for the Museum of Modern Art, is still very far away. In the meantime, she and the other foot soldiers of New York oyster restoration are pursuing approaches that are much more primitive and much more modest. With the oyster nursery established at the Harbor School and the oyster survivorship data tabulated from the city's oyster gardeners, the various nonprofits and volunteers committed to New York City oyster restoration chose six sites throughout the New York Bight and began building test reefs at those locations. The reefs are situated at the cardinal points of the oyster's vanished kingdom. Off Governors Island on the Bay Ridge Flats, up by the Tappan Zee Bridge near Hastings-on-Hudson, off Gravesend in Staten Island, in Jamaica Bay, and finally over in Soundview in the Bronx. All of this is coordinated by the Hudson River Foundation, the estuary's largest funder of scientific research, formed in the wake of the Clean Water Act's passage.

After visiting with Orff, I decided to take a look at one of these structures for myself. My guide was the Hudson River Foundation's oyster point person, Jim Lodge, a wry, pleasant man in his forties who seems as committed to his task as he is dismayed by it. A career scientist who has locked horns with nearly every city and state agency on behalf of the oyster, he routinely finds himself playing the pessimistic straight man amid all the oyster restoration euphoria. Driving with Lodge out to Soundview now, I glanced at him from time to time as he stared despairingly out the window at the urban ravages abutting the Brooklyn-Queens Expressway—tangles of housing projects, reams of pipes and wiring, the befouled Newtown Creek. "The goal is to put in five hundred acres of reef by 2025," he said out of the corner of his mouth. "But I mean, look at all this . . ." His words trailed off and he shook his head.

Arriving at Soundview, Lodge and I put on waders and began a slow shuffle out to the test reef. Around us, several dozen teenagers took part, some from the Harbor School, some from the Bronx River Alliance, still others from the Bronx-based group Rocking the Boat, an organization that teaches inner-city youth to build wooden boats from scratch, which they then use to navigate the Bronx River.

Out ahead of us, beneath the buoys, lodged in the seabed, was the oyster reef itself. At its base were several tons of rock that had been laid down the previous year by the Army Corps of Engineers. Atop that, clamshell had been dropped. And finally on top of that, a speckling of oyster juveniles called "spat-on-shell." Of all the reef structures that had been established in the New York Bight in the last two years, Soundview had been the most successful. Survivorship of juveniles had been strong, similar to the survivorship of oysters in cleaner waters.

But the project was at a standstill. The reef was a few hundred square meters—a meaningless amount relative to the restoration objectives. Lodge had been in discussions with government officials to expand the site, but after New Jersey had shut down the reef at Keyport, New York's Department of Environmental Conservation was reluctant to grant a permit. And it's not just human health fears that drive the worry around reintroduced oysters. There is also concern about the health of the oysters themselves.

Beginning in the mid-twentieth century, a series of oyster diseases started appearing on the East Coast. In the 1940s, the protozoan dermo migrated from the Gulf of Mexico and started showing up in the Northeast. For years it had remained dormant, since it appears to have evolved in the warmer waters of the Gulf of Mexico. But as climate change has warmed northern Atlantic waters and as oystermen increasingly planted dead beds with seed from the south, dermo spread with it.

A second, more deadly disease appeared on the East Coast in 1957, further complicating matters. Multinucleated sphere unknown, or MSX as it's called more commonly, showed up on the East Coast as the direct result of an oyster collapse that occurred on the West Coast. Just as industrial pollution and habitat destruction ruined wild beds in New York, gold prospecting along the California coast in the mid-nineteenth century released cascades of pollution and sediment into Northern California waters. Beds of the Olympia oyster, the only oyster native to the California coast, were devastated. In a desperate effort to replace the lost Olympias, growers began importing another oyster, *Crassostrea gigas*, commonly known as the Pacific oyster, from Japan. But *C. gigas* did not come to our shores alone. MSX traveled with it. And when Pacific oysters were introduced to Delaware Bay in an attempt to boost existing supplies of Eastern oysters, the disease wreaked havoc. Throughout the sixties, seventies, and eighties the already depleted Eastern oyster suffered wave after wave of disease collapse from MSX. It is the fear of these two diseases, MSX and dermo, that makes the Harbor School's idea of putting a "billion oysters" in New York Harbor frightening to public environmental officials. Even though New York Harbor is no longer a viable commercial oyster ground, diseases introduced into the harbor could conceivably spread northeast into Long Island Sound, where viable oyster farms still exist.

All of this troubles the environmental idealist. I, like others who have taken an interest in the New York oyster, wanted it all to go faster, to proceed more boldly. I wanted to see all the environmental bad deeds of the past somehow reversed by an ebullient resurrection of New York oyster reefs. But I saw very little of that as I trudged toward the nascent Soundview reef. I stared down into the water and muttered to myself about the stasis that had occurred between bureaucracy and bivalve, wondering whether the whole thing wasn't just some doomed high school experiment.

But as I trudged out toward the reef I noticed something flashing below on the surface of the mud. Taking shape below was a grayish form. It had nothing to do with the artificial reef and the reintroduced oysters farther down current. No, the thing at my feet required no introduction—it was a native shell. I reached down and tucked my fingers underneath its telltale bumpy exterior, expecting to run them along a hollowed-out recess, long ago emptied by some marauding human force. But my fingers found no concavity. Instead I felt a bottom shell concealing a live and viable organism. I lifted it from the water and felt the satisfying weight that any digger of bivalves recognizes as money in the bank.

Yes, the thing in my hand was a real, live, naturally spawned New York City oyster.

Holding it up to the light, I appraised it now and looked

at it for what it was. A brave sentry from a lost kingdom. And it was beautiful. It was what people in the oyster trade would have called "a single." With no reef to cling to, it had formed a perfectly shaped, free-tumbling shell that when scrubbed and cleaned would suggest the poesy of the sea without any of the gunk. It might have fetched upward of two bucks.

Jim Lodge spotted me standing in the Soundview waters staring at it.

"Oh, you found one. Let's take a look."

To my surprise, he quickly produced a pocket knife and found the creature's principal duct, located near its triangular tip. Sticking in the knife, he worked it sideways, gradually stretching out the adductor muscle until a little of the liquor slipped out, indicating a breach. He then slipped in the knife and cut the adductor on the oyster's ventral flat side. This, followed by a quick sweep of the knife over the interior of the dorsal shell, laid it bare in the summer sunshine.

I stared into the still-beating heart of the oyster and felt something strange: a mixed-up muddy feeling as murky as the Bronx River at my feet. I felt admiration and desire for this animal—this creature that had survived so much against such long odds. I felt anger at the people who had removed its relatives with such disrespect and at the same time I felt intrigue at the *merroir* this survivor held within itself. I wanted to rage at the Dutch colonists, who, instead of returning all those wonderful shells back to the water, where they could have made a home for so many other oysters and perpetuated New York's population, had instead burned them all down for lime, crushed them up into roadbed, discarded them, disregarded them. I felt disgust at the government bureaucrats who'd given up on the oyster, accused it, forcibly removed it, denied its right to exist—who had branded it with sterile legal terminology, called it a "nuisance," and decided it had no right to be in these waters.

Taking the oyster up again, I looked at it and slowly drew it to my lips. Its shell had already been breached. It would die soon. But right now it was alive and held within it the true energy of New York City. The essence of this place where I lived. A city official standing nearby whose many overtaxed duties probably included preventing New Yorkers from eating local oysters raised her eyebrows high and emitted a tiny gasp.

All reason abandoned, down the hatch went the New York oyster.

It was delicious, if a bit warm.

Sandy's Revenge

When we commit an act of transgression, we tend to anticipate immediate retribution. Some higher force punishing us from above or, worse, from inside. For a day I lay like Dostoevsky's Raskolnikov on my couch, contemplating the poached oyster in my belly. The old nineteenth-century Jewish American curse "You should get cholera!" boomed in my head. Would hepatitis turn me as yellow as a banana and leave me begging for a liver transplant? Or would I fall victim to neurotoxic shellfish poisoning and chew off my lower lip, as some victims of the disease have done. Or perhaps blissfully I would have my memory erased by amnesic shellfish poisoning—another common affliction that can arise from shellfish harvested in tainted waters. All of this was possible. Much of the pollution that had transformed New York waters from a food system into a waste disposal was still there.

But while no retribution came that day, nature was preparing a larger punishment—an attack on all New Yorkers for crimes against our estuary.

On October 29, 2012, Hurricane Sandy made landfall just off the southern tip of New Jersey and followed a path that marked very closely the ancestral footprint of the Eastern oyster. The convergence of a full moon would create a tidal surge that measured as high as fourteen feet, three feet higher than the eleven feet predicted.

All around the former oyster kingdom the damage would resonate. In Staten Island, where humanity had foolishly expanded into the historical intertidal—what really should have been the oyster's exclusive domain—destruction ruled. The sea topped the seawall of the artificially armored shoreline and then corroding salt water sat behind those seawalls instead of draining out and rotted away tens of millions of dollars worth of real estate.

To the south the sea would sweep over the Rockaways barrier island, merging with the waters of Jamaica Bay, short-circuiting electrical panels in hundreds of homes and setting them on fire from the communities of Auburn to Breezy Point. In the dredged-out, artificially deepened, and oysterless Jamaica Bay, the water would gain momentum and flood over the Belt Parkway and inundate shoreside homes from Sheepshead Bay to Howard Beach.

Farther into Brooklyn, the Gowanus Canal, that Superfund site where the oyster-tect Kate Orff planned on birthing oysters, a different kind of nativity took place. A toxic afterbirth of Gowanus sludge belched out all over the neighborhood. The sludge would sweep up and penetrate into the foothills of the gentrified neighborhoods of Park Slope on one side and Cobble Hill on the other.

In my own neighborhood in lower Manhattan, the dredged-out confluence of the East and Hudson Rivers surged up and flooded past the 1627 high-water mark and crept up the hill toward my home on the ridge crest of Broadway. On the morning Sandy hit Manhattan, my family and I holed up in our apartment while the wind blew. Amid all this I wrote an op-ed for the *New York Times* about the oyster, and how it once protected our shores and how it might once again.

The next morning I went out to try to find a hard copy of the *Times* to save for my files, but the neighborhood was closed for business. In search of a newspaper kiosk, I wandered farther and farther east, eventually ending up at the old Fulton Fish Market. Every swanky new bar and restaurant looked as if it had been victim of an angry brawl. Tables and chairs had been tossed around. Outside the Paris, one of the last restaurants that the old fishmongers used to frequent, a stream of full beer bottles ran out the door and poured out onto South Street. But the old Fulton Fish Market sustained hardly a lick of damage. It was a true waterfront building meant to withstand the washing in and the washing out of the tide. Even though it was empty now, it could have been used had someone wanted to. I couldn't help but think of what Kate Orff had said. We had to plan for a much wetter city in the future.

In the months to come, almost every waterfront planner would have a new idea about what to do about this wetter city. Money was in the air and everyone wanted a piece of it. And in spite of everything that Sandy should have taught us about how estuary systems work, how occupying the waterfront should be a game of give-and-take between sea and land, all the old forces that had dredged and filled so much of the New York waterfront were again clamoring to do the same thing all over again. From the Bloomberg administration came word of the idea to build something called Seaport City, a mirror of the sterile blocks of Battery Park City on the western side of the island. According to that plan, the shore would be armored even higher, topped with giant towers. In addition, large new amounts of land might even be created: a giant new dump-and-fill project that would eliminate the strait between the Fulton Market and Pete Malinowski's Harbor School, turning Governors Island from an island to the ignominious end of a rock-and-rubble, oyster-hostile peninsula.

But there were also murmurings in the oyster community about seizing some of that funding. For Jamaica Bay, a researcher from Stevens Institute of Technology named

Philip Orton put forth the idea that the bay be allowed to silt in naturally, to shallow. With a shallower and potentially oyster-laden bay, a storm surge coming from the south could potentially be reduced by the greater friction and drag that these features would exert on floodwaters. In the past the idea of limiting harbor dredging would have been abhorrent to the Army Corps of Engineers, the principal agency that sets the depths and determines the physical geography of the harbor. Now suddenly Orton was invited to publicly discuss the idea with the commanding officer of New York's Army Corps of Engineers in front of a gathering of New York's mayoral candidates. Kate Orff was invited by Mayor Bloomberg to join the post-Sandy rebuilding and resiliency initiative and was given a grant from the Department of Housing and Urban Development to start researching ecological infrastructure throughout the city. In Wallabout Basin at the Brooklyn Navy Yard, Baykeeper was given permission to open a second oyster nursery, and as of this writing 250,000 more oysters now hang from ropes in that most polluted bay, spreading their offspring far and wide. Even on the New Jersey side of the harbor, where Governor Christie had once banned oyster restoration, Baykeeper has been given permission once again to put in a larger reef (albeit at the highly security-controlled Naval Weapons Station Earle).

But the most significant thing to happen post-Sandy occurred at Soundview, where I had eaten that potentially contaminated oyster and the only place where I had actually seen wild oysters alive. There the oyster restorers got the go ahead to install a one-acre reef—the biggest New York City had seen in a hundred years.

At five thirty a.m. on a June morning in 2013, I swung by Brooklyn and picked up the Hudson River Foundation's Jim Lodge again and drove back out to Soundview. It was the low tide of the supermoon and hence one of the lowest tides of the year. Donning waders, we went out into the muck. The super low tide afforded us not only a view of the test reef that had been put out earlier but also all the miraculous "natural set" of oysters that had happened since I'd last been at the site. Hurricanes, while generally seen as destroyers of things, can for oysters be great boons. In the year following a hurricane, spat is blown far and wide and oftentimes a spectacular natural set occurs in a hurricane's wake.

This might just have happened at Soundview. Two hurricanes in a row—Irene and Sandy—appeared to have blown in a lot of spat. All around us the scores of old abandoned tires were now completely encrusted with new oysters. On some of the tires minireefs were forming. The Baykeeper Oyster Restoration program director, Meredith Comi, bent over to examine one of the clumps. "Vertical accretion, baby!" she exclaimed happily. The oyster-laden tires exemplified an interesting point. Humans can do all they want trying to dream up structures, build reefs, and introduce laboratory-reared larvae. But nature in the end is a lot less picky and a lot more random. What natural oysters really need is clean water and structure. And it seemed that these vital elements, forty years after the passage of the Clean Water Act, were hitting their stride here at Soundview. The previous year's Hurricane Irene may have hastened the process by blowing the available spat to the place where they could grow. Paradoxically, that place had been old radial tires. Jim Lodge shook his head at it all. "If we had a different regulatory environment," he muttered, "we'd just dump a bunch of tires in here."

But given today's difficult regulatory environment, in which the materials you can and cannot put into the water are strictly monitored, we were playing by the rule book and waiting for a lot of clamshell (cured for a year to remove all biological contaminants) to arrive by barge from Massachusetts. Hurricane Sandy had made any program having anything to do with the waterfront move exponentially faster through the bureaucracy, and suddenly Jim Lodge had the permission to do what he'd always wanted to do: build a reef that was bigger than just a science experiment. All 125 tons of that shell he'd just procured was about to arrive here in the Bronx at any minute.

The plan was to mark the muddy, oysterless areas in between the natural set of oysters and fill in those areas with clamshell. The hope was that by introducing enough shell between the areas of natural set Lodge would in effect connect the dots, making the isolated patches of oysters join together in a bivalvacious harmonic convergence.

Soon we spotted the barge *Arlene* rounding the corner. Lodge got a call on his cell. "I think I see you," he said, "unless there's another barge with a hundred tons of clamshell coming into the Bronx River." We rendezvoused with the boat and stepped aboard. The barge's commander, a lanky Israeli American named Gadi Zofi, sat down with Lodge and looked over a nautical chart. The barge drew five feet. At flood tide the reef area would be eight feet deep, but right now as the tide was rising it was only about three feet. We'd have to wait a bit and then slip in with the barge just as the tide was cresting, dump the shell, and get out before the tide fell too far.

"We don't do it quick," Zofi said, "we gonna spend the night in the Bronx."

Finally Lodge made the call to do the dump. The barge was detached from the tug and then a smaller guiding boat pushed us into the reef area. The engines roared and an oversized hose was hoisted above the pile of shell. With the area properly marked, the barge was "spudded down"—locked into place with a metal rod fixed to the bottom. And then the hydraulic hose spewed with the force of a dozen fire hoses into the shell pile. The first clump of shell slid off the barge and disappeared into the depths of the Bronx River.

Onshore, all the many people who had been working to bring oysters back to New York—the Harbor School students, the volunteers, the puny-salaried nonprofit workers, the harassed and underfunded Parks Department employees—all of them gave up a great cheer as the shell went in with a magnificent hiss.

The work would get frantic and pressed. Commander Zofi would start grabbing the supersized hose with his own two hands, desperate to get the shell off his deck and into the right place before the tide fell. But deposit the shell he did. The barge *Arlene* escaped a night in the Bronx, as did the rest of the oyster-restoring crew.

In a few more nights juvenile oysters raised at the Harbor School would be dropped on the cultch, hopefully making the Bronx their home.

Will these oysters ever be edible? Most say no, not in my lifetime anyway. But in a way that is beside the point. Long before New York was New York, it was the Hudson River estuary. Though we moved here and made our livelihoods here, did we have to cut out the city's seafood heart? Did we have to destroy twenty-one thousand of the twenty-five thousand acres of oyster-laden salt marsh our estuary once possessed? If you run the numbers, you find that this marsh would have been capable of producing all the seafood New York City could possibly need. This is true for waterfront communities all over our country. San Francisco, Seattle, New Orleans, Boston, Washington, D.C.—these are all cities built atop salt marshes and seafood nurseries. We have established dense populations in places of great biological wealth, decisions that have caused great damage to our marshland and the species that depend on that habitat. Is there not a way to share the intertidal, the place where oysters and humans meet?

As the New York oyster restorers suggest, maybe there is a different way of doing business with the waterfront. We're at a moment of unique opportunity. All over the United States a lot of money is about to be spent as sea levels rise and storm surges become more prevalent. In New York alone some $50 billion may be invested in the waterfront post-Sandy. Of course many people are salivating over the possibility of getting their hands on some of that money. There is no shortage of desire to move back into the intertidal, to rebuild things as they were, claim damages, get a share of the superstorm pie.

But, moving forward, maybe something else could be done with that $50 billion other than a simple patch and fix. Around New York discussions now percolate of a "Blueway," a girding of living, biologically friendly infrastructure that would serve as a new kind of edge to the terminus of the city. And it is high time for such a development, not only for New York but for every place humans build their settlements on the coast. The ocean is coming at us in a way it never has before, and very soon we will be forced to profoundly renegotiate a truce between land and sea. For millennia the typical way the Eastern Seaboard and the Atlantic drew up their treaty was by establishing a living, biologically dynamic border called a salt marsh, which was protective of the coast and productive of seafood. Underpinning it all was a miraculous natural architect called an oyster. Yes, we are a long way off from once again letting the oyster dictate the terms of the truce between land and sea. But the people who understand the importance of this vision are now graduating from places like New York's Harbor School. The next time land and sea clash in to hammer out terms, I would hope that these graduates are present at the negotiating table.

Shrimp

The colonists who settled our coasts in the seventeenth century were seafaring people. They partook of oysters and clams and roe and fish flesh of all kinds. But as their heirs pushed westward toward the Mississippi into the heart of the continent and developed this nation into a world-class landfood producer, the country drifted away from the ocean and toward beef and corn.

There was one exception to this trend: shrimp.

In the past century shrimp has gone from a sidebar curiosity sold mostly in ethnic markets to the very soul of our seafood economy. So thoroughly do shrimp dominate American seafood today that it is almost a menu category in and of itself—a type of seafood that people who generally don't like fish all that much will eat with relish. A decade ago shrimp surpassed canned tuna as the most popular seafood in the United States and now the average American eats more than four pounds of it a year—roughly equivalent to the U.S. per capita consumption of the next two most popular seafoods—tuna and salmon—combined. If they didn't eat shrimp, most Americans today wouldn't eat any seafood at all.

The switch to shrimp isn't just a random redirection in shopping patterns like choosing hamburgers over hot dogs. It is a paradigm shift. If oysters tell the story of the very first local seafood to disappear, shrimp tell the story of the unraveling of the entire American seafood economy. Fifty years ago, 70 percent of our shrimp was wild and much of it hailed from the Gulf of Mexico. Today 90 percent of our shrimp is farmed and imported, mostly from China, Thailand, Vietnam, India, Indonesia, and Ecuador. Many other fish and shellfish that Americans consume are trending in the same direction.

But it all began with shrimp, a product that slipped into our domestic seafood infrastructure and opened up our markets to Asia. This occurrence has effectively decoupled the American consumer from the American coast and reattached us to ever more distant shores.

The current shrimp market is the result of the interplay between two dominant forms of shrimp acquisition: American fishing and Asian farming. The parallel tracks become increasingly intertwined as Asian and American traditions have overlapped, quarreled, and resolved themselves over the last 150 years. There are many places to begin the shrimp story—in the now mostly defunct shrimping grounds of San Francisco Bay, in the floodplains of Asia's vanishing mangrove forests. But perhaps it's best to start in the place where the shrimp fight is most pressingly linked with America's ability to feed itself from the sea. That place is the mouth of the Mississippi in the state of Louisiana, the greatest shrimp-rearing ground in North America and the site, incidentally, of the biggest oil spill in U.S. history.

A Drag in the Dark

"This year the boys didn't eat the first shrimp of the year, raw with olive oil and sea salt, as they usually do," a Louisiana shrimper I'll call Denise Aubrey wrote in an e-mail in the summer of 2010, a month after the BP oil blowout began. "The trawl was full of tar balls. The oil has reached our sweet brown shrimp in Lake Pontchartrain. The catch had to be thrown back. The factories have shut down . . . We will attempt as always through sickness, hurricanes, and oil spills to preserve our heritage and fish off this beautiful land. We will prevail."

A year after receiving this upsetting communication, I

traveled down to the shores of Lake Pontchartrain, a shallow brackish body of water that skirts the edge of the Mississippi delta and leads out into the Gulf of Mexico. I wanted to see what had happened to that shrimper since the spill. Were she and her husband shrimping? Had things recovered? Could you eat the shrimp from the Gulf? These were the questions that seemed on everyone's mind at the time. But as I spent an evening fishing with the Aubreys, I came to understand that the story of shrimp in the Gulf was much bigger than the spill. In fact, shrimping with the Aubreys revealed to me how all of American fishing had changed over the last century and how oil and other big extractive industries had been undercutting American fishing for decades.

As I stepped aboard his boat, Denise's husband, whom I'll call Cal Aubrey, assured me that, in spite of the spill, things were going to be all right. "This resource is so resilient, it's just incredible," he said. In every corner of the commercial dock there was real activity and it did indeed seem, in the summer of 2011, a year after the BP blowout had been capped, that things were coming back to life. It was about the most iconic American fishing scene you could imagine. Shrimping boats of different sizes geared up, their attractive green "elephant ear" nets billowing in the light breeze. A slightly sweet and rank smell of crab and shrimp husks discarded by the dockside processors was held close in the hot June air. Aubrey's youngest son, a sleepy boy clearly enamored of his father, boarded with us, intent upon staying up for the night. The boat's two Latin American crew members donned rubber boots and busied themselves with the large orange hampers that would be used to sort the catch and tended to a payload of ice that would cool it all down when it came out of the water that, already in June, was as warm as stew.

As the deckhands prepared to cast off, Aubrey's son started to fade and Cal put him to sleep belowdecks. Shortly thereafter we left the dock and headed out beneath an orange sky, skirting the remnants of the Louisiana marsh, the heart and lungs of American shrimp country. For miles and miles the marsh seemed to stretch, an undulating primeval ecosystem that is as essential to American seafood today as the oyster reefs of New York Harbor were to the seafood of yesteryear.

Aubrey hails from a family that traces its residence in Louisiana's muddy boot back two hundred years. His cousin was an inveterate fisherman, and when Cal was in his late teens, in the mid-1970s, he drew Cal in. "My cousin's parents wanted him to go to college," Cal told me. "Well, he did. He got a degree and then he came home. And then he gave his mama his diploma. And then he went fishing. And he took me along with him." From the moment Cal joined his cousin it seemed like a good bet. Cal's tenure in the Louisiana shrimp fishery paralleled exactly the great Gulf shrimp boom (and indeed the great American fishing boom) that started in the 1970s and continued until the boom busted in the 1990s.

But two hundred years ago, when the Aubreys first made landfall in Louisiana, it's unlikely that anyone could have made a living off of shrimp alone. When our nation first started to encounter shrimp, only a minor market existed for them among American diners of European descent. Whatever was caught locally in Louisiana was mostly consumed locally. It was only when Chinese immigrants encountered shrimp in California that the 150-year process of turning shrimp from a local staple into a global commodity began in earnest.

As Jack and Ann Rudloe explain in their book *Shrimp: The Endless Quest for Pink Gold*, the Chinese American shrimp industry began when Cantonese laborers toiling in the gold mines of Northern California started going fishing. After the gold rush, some émigrés who managed to find work in other industries discovered large populations of *Crangon franciscorum*, or bay shrimp, on the shores of San Francisco Bay. A small Italian fishery of eight boats existed at the time, but once the Chinese began employing shrimp-seining technologies from their home coastal provinces in the Pearl River delta, the shrimping industry expanded significantly.

Just as New York Harbor was once a rich, productive estuary replete with oysters, San Francisco Bay was thick with shrimp, thanks to the verdant salt marshes surrounding the fertile San Joaquin and Sacramento River deltas. Many of the two thousand species of shrimp in the world could not exist, in fact, were it not for salt marshes. After hatching in the open sea, most commercially valuable shrimp make a beeline for the coast. Once inshore, they seek out protective cover to evade predators. A salt marsh is the ideal place. Not only do marshes offer nooks and crannies for juvenile shrimp to hide, they also provide the crustacean equivalent of mother's milk. Shrimp are highly omnivorous and as juveniles subsist off decomposing swamp muck that slips off the edge of the marsh into the murky waters.

It was exactly this kind of environment that made San Francisco Bay so productive for shrimpers in the mid-nineteenth century. Hundreds of millions of pounds were harvested. A few were sold fresh to white settlers as "San Francisco cocktails." The majority were exported to China.

The preservation method of choice was air drying, which turned them into a transportable commodity that could be kept in a ship's hold for weeks during passage to Canton.

But in just a few decades, the San Francisco Bay shrimp fishery was in decline and the marshes upon which the shrimp depended were being destroyed directly and indirectly by the gold industry—the same gold industry that had laid waste to California's Olympia oyster beds. Hydraulic gold mining washed immense amounts of sediment down the rivers into the bay, creating new areas of mudflats in place of salt marsh. Washing down with the sediment were toxic materials involved in gold processing, including mercury and arsenic. New settlers to the area also filled in marshes to make farmland. And still another destructive industry, salt making, started to use the bay. The salt industry further degraded marshland by causing brackish waters to become overly salinated, a process that can kill many species of marsh grass as well as other wildlife. Once marsh grass dies, a coastal estuary begins to fall apart and ceases to be the vital nursery it should be. By the mid-twentieth century, the marshes of the bay had been greatly reduced, and with them, mainly due to declining demand, the shrimp industry also faded. San Francisco Bay is still suffering the effects of the nineteenth-century gold rush and salt poisoning to this day. The same wetlands impingements felt in New

York—i.e., diking, filling, and dredging—also continue to hamper the ecosystem's recovery. When Europeans arrived in San Francisco Bay, it is estimated the region had over five hundred thousand acres of tidal marshes. Today there are less than forty thousand acres.

During the same period that San Francisco's salt marshes began going into retreat, a similar assault on seafood-producing salt marshes was occurring throughout the United States. Beginning in 1849, the federal government passed a series of reclamation laws called the Swamp Land Acts. These acts basically subsidized the destruction of salt marsh, using federal money to underwrite the drainage of wetlands all around the country so that those lands might be repurposed to settlement and the production of land-food. The Great Dismal Swamp of Virginia and North Carolina, the Okefenokee Swamp of Florida, and the Cacaw Swamp of South Carolina were all significantly drained during this time. Unbeknown to them, Americans were also draining away their seafood in the process.

But there was one place that had so much marsh that even the Swamp Land Acts could not dent it: the marshes surrounding the Louisiana portion of the Mississippi delta, the largest marshland in the continental United States. It was here that the Chinese and other cultures alighted in the next phase of American shrimping.

Through the introduction of drying technology, the Chinese had laid the essential groundwork for an American shrimp economy. They'd showed that shrimp could be preserved, turned into a shelf-stable product, and in turn shipped overseas, potentially becoming a valuable export commodity. And so while the San Francisco Bay shrimp fishery failed, the Chinese shrimp trade was able to keep going in southern Louisiana.

Both geographically and ethnically, Louisiana is a constantly shifting state—one whose very physical shape changes with storms and tides, and whose land has been colonized and recolonized by diverse waves of immigration. In geologic terms it is very young, formed by muddy alluvial deposits brought down by the Mississippi River. Its immense, tremendously productive marshland hosts not only shrimp but an abundance of commercial species, ranging from the largest wild oyster reefs remaining in the United States to prolific schools of redfish, black drum, and many others. If you wanted to have a vision of the productivity of the Hudson River estuary two hundred years ago, a visit to the still abundant and diverse fisheries that surround the Mississippi delta would be a good start.

Culturally, Louisiana is also a wild, mixed-up place, where individuals or entire ethnic enclaves could disappear into the marsh and do their own thing. Whereas the Chinese who arrived in San Francisco were for the most part the chattel of the gold and railroad industries, the Chinese who made their way to Louisiana were able to more freely strike out on their own.

Louisiana therefore became a destination for nineteenth-century Chinese fishermen looking to get a foothold. Literally. As University of Louisiana historian Carl Brasseaux has noted, in the late 1800s Chinese immigrants lived and worked in the inhospitable wetlands that were considered no-man's-lands by less tenacious people. Building their houses on stilts, they surrounded them by corrugated wooden platforms anywhere from one to three acres in size. After harvesting shrimp they would put out their catch, at which point other townspeople were invited to come and "dance the shrimp." "The people would come along and shuffle their feet over the shrimp to remove the shells," Brasseaux explained in a recent interview. The winds would blow away the shells and then the dried shrimp were cleaned and sent to Chinatowns all across America and all over the world.

Non-Asian Americans remained little interested in dried shrimp, and so, outside of Louisiana, shrimp remained a niche product confined to ethnic markets. But within the state, shrimp became a mainstay of a much older Louisiana group—the French-speaking Cajuns of the coastal bayous.

Like the Chinese, the American Cajuns were an exile people, driven first out of France, then Atlantic Canada (the word "Cajun" is a colloquial corruption of the name of the Canadian territory Acadia), and finally arriving much harried in Louisiana, where they subsisted off the Louisiana marshes. The Cajun eating tradition has historically been a rough-hewn one marked by extreme abundance—one where diners roll up their sleeves and pull whole sea creatures apart with their bare hands. Coastal brown shrimp were just one of many difficult-but-plentiful-to-eat things Cajuns enjoyed. Oysters, crawfish, biganot snails, alligators, even a now largely unknown river prawn that had once migrated from the Mississippi's mouth all the way north to the Ohio River—all of these hard-shelled creatures have historically been the Cajuns' labor-intensive daily bread.

To this day an extremely high level of seafood consumption sets the Cajuns apart from the rest of the country. In the brackish sluiceways and milk-tea bayous of the rural part of the state, the standard unit of measurement for local seafood is the sack. A sack is a burlap bag that can hold roughly six dozen oysters, twenty dozen crawfish, or thirty dozen biganot snails. Any of these creatures might end up in a sack at any given month of the fishing season, and it is not unheard of for a family in the bayous to go through a sack in a single day several times a week. Shrimp was a unifying part of this potpourri, a food that brought together both poor swamp people and refined city dwellers. As a pamphlet produced in the 1950s by the U.S. Fish and Wildlife Service marveled, shrimp in Louisiana "are a mainstay food in all homes—mansion and cabin, plantation house and trapper's shack, city apartment and fisherman's houseboat. They are served economically and simply as in stew, gumbo or jambalaya or elegantly as in the meunière and amandine of old French restaurants."

This Louisiana celebration of seafood abundance was something that started to burst into American culture at large only in the twentieth century. And the most entrepreneurial members of the Cajun community began to try to figure out how they might capitalize on that appeal and reach a larger consumer base.

Shrimp presented an attractive but confounding puzzle in this respect. Unlike fish, which have internal skeletons and a loose, light musculature, shrimp sport heavy armature that directly affects the quality of their flesh. In the course of their lives shrimp regularly shed their shell in several different molts. Each time they molt they generate a new shell from a dense layer of collagen in the peripheral layer of their flesh. The denseness of the collagen shell precursor makes shrimp flesh have a porklike mouthfeel that is distinctly different from fish—a quality pleasing to those who are more disposed to eating meat.

That same heavy exoskeleton makes shrimp resistant to automated processing. In the words of the shrimp-pushing promotional pamphlet quoted above, "The armored heavy-headed little fellow has posed problems of supply and distribution as thorny as himself." But the pamphlet continued, "New Orleanians, like the old Creole masters of cuisine, have been creative and very busy. Their aim has been to spread the shrimp supply to inland markets so that women everywhere could buy them easily."

The creative and busy Cajun who cracked the code and made shrimp available to Americans everywhere was a man named J. M. Lapeyre of Houma, Louisiana. A polymath credited with inventions that ranged from a digital magnetic compass to a jet engine, Lapeyre started out as one of eight children in a modest shrimping family. His transformation from factory worker to inventor began one day in 1943 when he noticed how a shrimp rubbed by his rubber boot slipped from its shell. Burning with the idea of a series of rubber rollers that would, like his boot, grip a shrimp's shell and extract the meat, he tried out his idea on his mother's hand-cranked washing machine and found that it worked. Then with his father and uncle, he successfully designed the first shrimp-peeling machine, which would turn Lapeyre into a multimillionaire. At the time of the peeler's invention he was just eighteen years old.

Lapeyre's processor moved shrimp steadily forward into the American consciousness. The next piece in the puzzle was a means of preserving all that processed shrimp. From the days of the Chinese drying platform in the 1800s, techniques advanced with the advent of canning. Napoléon Bonaparte had got canning going in 1795 by offering a twelve-thousand-franc prize to anyone who could arrive at a way of preserving meat for soldiers in the field. Thereafter, in 1812, the year Napoléon's field army disintegrated in the snows of Russia, the Americans Ezra Daggett and Thomas Kensett produced the first canned seafood, a New York oyster. They also canned salmon and lobster. But the quality of canned shrimp never carried the same appeal as the fresh product. The thick, dense flesh of shrimp became mealy.

Freezing changed all that. Industrially freezing seafood emerged in the late 1920s when Clarence Birdseye invented the "double belt" freezer that froze seafood quickly. Before the double belt, seafood could be frozen only over a period of hours. With such a slow freezing period, ice crystals would form within the animals' muscle tissue, breaking cell membranes and rendering the meat flaccid and mushy. With Birdseye's invention, ice crystals didn't have time to form, and shrimp's proper mouthfeel was at last better preserved. And as the years passed, freezing technology grew even more efficient and robust. By the 1960s temperatures of minus forty degrees became easily achievable, markedly increasing the shelf life of seafood. These developments were to aid the trade and transportability not only of shrimp but indeed of *all* American seafood. New England cod, California halibut, Alaska salmon, Hawaiian tuna—all of it now was ripe for a larger market. And with the building of the country's interstate highway system that was launched in the 1950s, seafood, once the most perishable and local of commodities, had become by the 1970s something that could be shipped virtually anywhere in the country.

In addition to the freezing, processing, and transport, there was a fourth part to a great convergence of circumstances that led to a significant upsurge in American fishing: a successful American grab for more ocean. For the first 350 years of New World maritime history, the ocean couldn't really be owned at all. During that period the seas were governed by the tenets of the 1609 treatise *Mare Liberum*, or *The Free Sea*. Drafted by the Dutch legal scholar and philosopher Hugo de Groot, *Mare Liberum* posed, in opposition to the then ocean-monopolizing Portuguese and Spanish empires, that the ocean should essentially be free for all. Restrictions on marine commerce, De Groot argued, hurt all nations, and all nations would profit should restrictions on using the high seas be lifted. After *Mare Liberum* was published and the Dutch asserted their freewheeling spirit in the global ocean trade, other nations started going along with the free-sea concept. As a result it became generally accepted that a nation should own the ocean only as far out as a cannonball could fly. This evolved into a global standard distance of three nautical miles from shore—a standard that held until the end of World War II.

By the time the 1970s arrived American fishermen had become increasingly impatient with the liberties the world's major fishing powers were taking based on *Mare Liberum*. They came to realize that while this centuries-old piece of wisdom may be good for global trade, it was rotten for American fisheries. The most productive ocean territory fishwise is typically the area from shore out to the continental shelf. With all of that territory essentially free for all, foreign factory vessels from as far away as the Soviet Union regularly set up shop within sight of American shores.

This phenomenon became a patriotic rallying point not just in the United States but around the world. As more and more nations felt the oppression of the largest and most aggressive fishing fleets, the international community gradually came to a consensus that the times of *Mare Liberum* had passed and that a privatization of the seas was needed. In

1976 the United States was among the first to take action. It passed the Magnuson-Stevens Act, in which it unilaterally declared a two-hundred-mile exclusive economic zone. This action would come into effect in 1983.

With foreign fleets hounded from U.S. waters, the American federal government took it upon itself to build up American fishing efforts to capitalize on all that newly available fish. In the 1980s tax incentives and government assistance programs to lower the burden of investment in fisheries were put into place. Upgrade followed upgrade. The American fishing fleet swelled to an unprecedented size. And with it, so did the American catch.

All of this brought prosperity to people like Cal Aubrey, the shrimper who was now guiding me in his boat down the edge of the bayou trying to cobble together a night's catch.

"I started out when I was fifteen years old with a little bitty boat," Aubrey told me as he got ready to put down his nets for the first shrimp drag of the evening. "I made enough money to buy a bigger boat. Then I wanted to sell that and I started building this boat." From boat to boat, his catch increased, as did his family's well-being.

But while Aubrey's catches grew, the specificity and localness of American seafood culture started to wane. Throughout the 1970s and 1980s seafood increasingly became a cheap, faceless commodity. The freezing and transportability of seafood meant that one kind of seafood from one ecosystem could easily be slotted in for another—so much so that "seafood fraud" became common, with anywhere from 25 to 75 percent of American seafood being regularly mislabeled. And with that facelessness also came franchising. Franchise restaurants devoted solely to seafood first rose to prominence in the late 1960s and early 1970s and took advantage of the uptick in transferability and supply. Red Lobster, spawned in Lakeland, Florida, in 1968 and later sold to the food giant General Mills in the 1970s, expanded exponentially in the 1980s. Another chain, Long John Silver's, was born in 1969 and expanded similarly. I recall the day that a short-lived regional restaurant chain known as Beefsteak Charlie's arrived near my hometown and announced an unlimited shrimp bar. The cheesy seventies advertisement showed a couple remorselessly attacking the shrimp bar, chanting, "Shrimp, shrimp, shrimp, shrimp, shrimp." So bloated with shrimp were they by the time their smiling waiter brought their main course that they looked at their hunk of beef in dismay and said their only other line of the commercial: "Steak?"

Similar things were happening around the United States. Just as the shrimp fleet grew, other fleets around the country also swelled, and with that growth came a surge in the exploitation of the fish they pursued. Initially it had been hoped that the Americanization of American waters would lead to better conservation and healthier fish stocks. And initially there was indeed more fish and shellfish to go around. But in the absence of serious regulation, combined with the ever increasing efficiency of fishing technology, America overshot the mark and overfishing became a serious problem. Restaurant chains and wholesale consortia consolidated fish buying. Prices went down. Fishermen overbuilt to make up for loss in per unit value of seafood. They overcaught and in some notable cases, like the codfish grounds of Georges Bank and the rockfish grounds of the California coast, outstripped the American ocean's ability to restock itself. As the 1970s progressed into the 1980s the American seafood deficit started to mount in earnest.

By the time I went shrimping with Cal Aubrey, it was clear that the boom created by this buildup was over. After our first long drag down the shoreline, Aubrey told his crew to haul in the two fifteen-foot-wide elephant ear nets. The nets came over the gunnels and were led over by the crew, who guided the catch to two plastic hampers. It was a sparse catch of shrimp—probably less than fifty pounds in all. Mixed in were baby flounder the size of sand dollars, spotted sea trout, catfish, and various other edible species. The fish were sorted and dumped back overboard. Some of the catch was alive, some of it dead, much of it easy prey for the various creatures that waited below the surface for an easy meal.

This kind of accidental death made environmentalists particularly angry, and the rage grew on many scores as American fleets built up in the 1980s. By the 1990s shrimpers were hit by a series of lawsuits for the accidental killing of endangered sea turtles. Now all shrimpers fishing in federal waters must carry a turtle excluder device, or TED, a kind of escape hatch that when operated properly should shunt turtles out of a shrimp net. In addition, sport fishermen grew upset by the fact that shrimpers often killed ten pounds of other fish for every pound of shrimp they kept. The popular American red snapper was particularly affected by shrimp bycatch and became something of a poster child for the shrimper vs. sport fishermen conflict. Soon after the sea turtle campaign was successfully waged, environmental and sport fishing groups began pressuring regulators into instituting requirements that boats trawl with bycatch reduction devices, or BRDs. And while both TEDs and BRDs markedly reduce shrimp catches, some more conscientious shrimp captains, like Cal Aubrey, have grudgingly come to accept them. "People need to realize that we don't wanna catch that shit either," Aubrey told me as the nets were cleared of the remaining bycatch for another drag. "It's just extra work. There's a ton of benefits to these bycatch reduction devices. If I'm gonna drag a bit, I'm not gonna want to drag a pile of extra fish around." Meanwhile, foreign shrimp trawlers fishing outside of American waters have different, often less stringent, rules for bycatch mitigation. They can be inspected for turtle killing by the U.S. Coast Guard, but the Coast Guard must ask official permission to inspect the contents of foreign ships' holds.

In addition to greater regulation of fishing practices, the 1990s saw a marked reduction in the numbers of fish fishermen could catch. In 1996 the Sustainable Fisheries Act was passed to supplement the initial legislation that had prohibited foreign vessels from American waters. For the first time in American history, this law mandated that all fish and shellfish stocks be scientifically and systematically assessed. In the process the act identified numerous stocks that had been overfished and mandated the rebuilding of every American commercial fish stock. Much stricter quotas were imposed and the buildup of American fishing that had taken place in the 1970s and 1980s started to ebb.

All this talk of fleet reductions, quotas, and bycatch made Cal Aubrey weary and jittery. In addition to regulatory changes, Hurricane Katrina (which sunk one of his boats and destroyed his home). Aubrey has had to marshal his strength against cancer. Aubrey lives in a section of Louisiana dubbed "Cancer Alley," a place where the oil industry has so polluted the air and water that it now sports the highest rates of cancer in the nation. And though Aubrey ultimately beat his disease, he has serious money concerns and throughout the night he was troubled about his haul. By four a.m. it was looking like a poor night of fishing. Haul after haul failed to produce what he calls a "Forrest Gump moment." Finally, after skimming shrimp all night long and burning many gallons of diesel, Aubrey's boat limped toward shore at dawn.

Once back in port, we encountered a traffic jam as a dozen-odd shrimping boats and sleepless, bleary-eyed captains elbowed their way to get to the entryway to the processor. They all were selling shrimp. No one was out pursuing the abundant black drum that swim in the near-shore waters. No one was chasing down flounder or sheepshead porgy or any of the other dozens of varieties of fish that could conceivably feed a family—those fish didn't fit into the Red Lobster commodity model. Since the shrimping boom started in the 1970s and Americans became obsessively focused on shrimp and shrimp alone, Gulf fishermen have grown progressively focused on delivering not a variety of fish for their neighbors but a commodity crop for a national market.

The managers of the onshore processing plant, with their

shrimp peelers, Birdseye-invented belt freezers, and wholesale delivery trucks ready to send seafood to the next link in the processed-food wholesale chain, were working the telephones with vendors around the country. This was a system that worked acceptably for everybody during the boom years of the seventies and eighties, when the seafood industry was expanding. But today shrimp is sold everywhere, easily frozen and traded, and scores of middlemen take their cut. The downward price pressure on fishermen is extremely high. And now, with the industry in retreat, fishermen are losing their foothold on the coast. Coastal land and dock space are being bought up for vacation and retirement developments. Increasingly, seafood processors can't afford to be on the sea. As Vito Gialcone, a cod fisherman in Gloucester put it to me, "Fish houses are getting turned into hotels all the time. But you never hear about a hotel getting turned into a fish house."

The fish houses that remain face little competition. Fishermen have no other choice, no opportunity to negotiate a better deal, and since by and large shrimp is all they go after there is no other alternative market for other species they might catch. They must accept the price the fish house offers.

This process was on full display as an extractor was positioned over Aubrey's hold. Shrimp soon began raining down in a pink waterfall. The much-hated "counter," a skinny harassed-looking man, emerged. Nervously, he scooped up a sample of the shrimp. Aubrey watched him carefully, almost trying to goose his hands over to the larger shrimp. The counter's job is to determine how many shrimp are in a single pound of a given sample. The larger the shrimp, the higher the price per pound, so a bad count can ruin a shrimper's day. Captains tend to circle around the counter like gamblers keeping tabs on a crooked blackjack dealer.

After checking the scale, the counter muttered, "Sixty-eight"—the number of shrimp in a single pound. It was a bad number not worth much money—the higher the count, the smaller the shrimp, and the lower their per pound price. A count of, say, twenty-five would be what the American consumer would call jumbo shrimp. Aubrey winced at the cursed number sixty-eight. His catch was worth about a dollar a pound. In all, his take for the night was around eight hundred dollars, barely enough to cover the cost of fuel.

And fishermen like Aubrey were not the only ones feeling the pinch. In the wake of the BP oil spill even the giant shrimp processors and packagers—the people who have squeezed prices so assiduously—were themselves being squeezed by a market fearful of anything from the Gulf. Everywhere I went in that summer of 2011, whether it was on the smaller docks around Lake Pontchartrain or the massive shrimp port on Grand Isle, shrimpers and brokers were scalding mad about the damage the spill had done to their product's reputation.

"What we've got is a bad perception problem," Louisiana's biggest shrimp broker, Dean Blanchard, told me on his dock in Grand Isle a few days after my night out with Aubrey. "We *know* we have the best shrimp in the world, but since the spill it's been a bad perception. You got 59 percent of the people in this country that don't want to eat this shit." Out-of-state consumption of Louisiana seafood dropped markedly in the year following the spill. In an attempt to put a Band-Aid on the hemorrhage, BP flooded Louisiana with $11.8 billion of payouts that baffled fisherman and processor alike in their randomness. Blanchard, an oft-quoted figure during the spill, who employs some of the more colorful languages I've encountered in the fishing industry, was overflowing with bile at the ridiculousness of it all. For weeks this wild-eyed Cajun who has moved millions of dollars of shrimp over the course of his career had been getting job-retraining letters from a place called Career Solutions in New Orleans.

"They asked me to fill out a form. Asked me what other career I wanted to do." Here Blanchard smiled mischievously and raised his eyebrows high. "You know what I told them? I said I wanna be a vagina piercer. What the fuck am I gonna do at fifty-one years old? I'm gonna be a designer vagina piercer. Instead of having my name on the back of some girl's jeans, I'm gonna have my name tattooed on their vaginas. Those Career Solutions people never wrote me no letters back."

All the free-market forces that entrepreneurs and hedge funders tell us raise all boats and "increase the fluidity of capital" do nothing of the sort for Dean Blanchard and even less for a smaller player like Cal Aubrey. The prices go down even if the shrimp catch declines.

And as embattled as Gulf shrimpers would already be with the fall in prices, increased regulation, and the BP spill, they are being pushed even further underwater by another pressure coming from far away, one that has taken advantage of all these setbacks. Cal Aubrey likes to sing the praise of the specific qualities of *Farfantepenaeus aztecus*— his beloved brown shrimp that thrives in the near-shore waters of the Louisiana marshes. "Our shrimp has a deep, rich flavor. These brownies, they have almost a sweet taste. You get a certain flavor from the wild. It can't be replicated. It's just far superior."

But it turns out that the modern American palate can't appreciate the specific nature of "brownies." Today, thanks to a revolution in food production, Americans would just as soon eat a shrimp from Cal Aubrey as they would one from the other side of the world.

Into the Mangroves

To get a sense of how profoundly shrimp have sculpted the modern geography of our planet, I recommend a flight to Ca Mau, the most southerly city in the Socialist Republic of Vietnam. I took such a flight in the spring of 2008 as I sought to trace the other thread of the shrimp story, the one that wends its way through the tropics of Southeast Asia. Departing from Ho Chi Minh City, I embarked on a creaky internal flight, due south. As we approached our destination, the pilot employed the typically communist approach to landing, dashing down abruptly rather than the slow, gradual descent favored by Western pilots.

Zooming through the clouds, I was thankful at last to catch sight of the shoreline. But as I looked closer I came to realize that the shoreline was not altogether there. It had been reduced to a smudgy dotted border. What was once a firmly defined natural hedgerow of mangrove forests was now pocked by a sweep of shrimp ponds, the result of a decades-long shrimp-farming juggernaut that is only now starting to ebb.

In the humble airport of Ca Mau, a small, gentle woman in her forties named Suong waved her hand cheerily to me. We had met via e-mail through the introduction of a French development worker, and Suong had efficiently looked up my photo online and easily picked me out of the crowd. A gaggle of American shrimp aquaculture researchers from Florida had also arrived on the same flight, but Suong dispatched them to another car. Me, she quickly whisked out of the airport and transferred to a clunky old motorboat with a heavy, hundred-plus-horsepower engine angling us scarily toward the vertical. Then eastward we flew up the Ganh Hao River—a bayou as muggy and mud-luscious as any Louisiana could boast.

The rotten smell of the durian fruit hung in the air. Slender wooden dugout canoes with long outboard engine shafts that cleverly tip up to avoid the rafts of floating vegetation buzzed past us toting produce back to Ca Mau. Some of the dugouts were lashed together into extemporaneous floating markets and manned by merchants trading frantically. The passing traveler could tell which food each skiff was selling because the food itself was hung atop ten-foot-high posts the salespeople had planted beside their boats. Though literacy in Vietnam has gone from only 10 percent in 1945 to over 90 percent today, out on the river a piece of food hung from a stick was a much more effective advertisement than any kind of printed sign. Purple banana flowers were slung over the posts, as were sweet potatoes and mangosteens. But amid this cacophony of food and commerce there was one thing that was notably underrepresented in the market hampers: shrimp. Seven out of every ten shrimp grown in Vietnam are sent abroad.

Before Vietnam's equivalent of perestroika, the so-called *doi moi*, or renovation, that opened up the country to foreign markets in the 1980s, this would not have been the case. For most of Vietnamese history, shrimp was a low-volume local crop, part of a long-standing multicultural tradition upon which many rural families subsisted. In fact, small-scale local shrimp farming is older than the nation of Vietnam itself. Asian societies, thousands of years older than our own, transitioned to shrimp farming long ago. It began when wild shrimp juveniles migrated into tidal impoundments that were built to entrap a favored product at the time, something called a milkfish. Shrimp accidentally wandered into these coastal ponds and ended up growing alongside the milkfish, eventually resulting in crops of 90 to 175 pounds of shrimp per acre per year. These ancient systems, referred to in aquaculture circles as "extensive" operations, became increasingly dedicated to shrimp over time. They produced shrimp in a manner halfway between farming and fishing. Channels were cut between natural inshore brackish ponds and rivers, allowing the flow of water to bring shrimp larvae into enclosures on an incoming tide. The enclosures could then be closed off and guarded until larvae reached a harvestable size. Because shrimp could easily feed off decaying organic matter (mangrove leaves falling into the water, for example), no additional feed was required. Because the densities of shrimp in these systems were low, disease outbreaks were rare. For centuries this had been a simple, relatively harmless way of acquiring a few grams of protein for families up and down the country's waterways. Not a lot of shrimp were produced, but enough.

But as Jim Carrier explained in a 2009 article in the journal *Orion*, when a certain specialty dish emerged that required tableside access to *living* shrimp all that would change. "Dancing shrimp," or *odori ebi* in Japanese, draws on a Japanese tradition called *ikizukuri*—a method of consuming sea creatures while they are still alive. In *odori ebi* a prawn is snatched from an aquarium, dipped in saki to anesthetize it, and cleaned and cut so rapidly that it still quivers when it reaches the diner's mouth—its antennae and legs often still "dancing" as it is consumed. Dancing shrimp was first popularized in prewar Japan and was made from a hard-to-come-by species called the kuruma prawn—an animal that when sold live retails today for as much as a hundred dollars a pound.

The idea to bring the kuruma prawn dance closer to home became the particular obsession of a young Japanese marine biologist named Motosaku Fujinaga just before the outbreak of the Second World War. Fujinaga came to shrimp at a time when his country was an economically polarized society, a place where the wealthy might have a chance to sample kuruma prawns but where commoners were not so lucky. Animal protein was a key to equalizing societies' nutritional imbalances. But in a mountainous archipelago nation where arable land was limited and terrestrially produced animal protein expensive, something that would efficiently utilize the limited space Japan had to offer would clearly be a boon.

Shrimp must have seemed to Fujinaga like a domesticated animal with great potential. To begin with, shrimp are highly adaptable to a wide range of feed. When they migrate to inshore waters, they may subsist on the most elemental forms of feed. Secondly, shrimp grow extremely quickly. In nature shrimp are an "annual" crop, going from egg to young male and then changing sex to spawning female in less than a year. Lastly, shrimp tend toward dense aggregations, schools so thick that in the heyday of Gulf shrimping it was not unusual for a shrimper to land tons in a single evening. This implied that shrimp raised on a farm could make intense use of just a little bit of space. Since a lack of land for agriculture was a key food security risk for Japan, shrimp must have seemed a perfect way to address that problem.

But Fujinaga was also an internationalist of the highest order, a man who, contrary to the rising nationalist sentiments of prewar Japan, believed that international problems must be solved multilaterally. It was for this reason that instead of writing his dissertation in Japanese he took the heretical risk of publishing in English. His thesis, "Reproduction, Development and Rearing of *Litopenaeus japonicus* Bate," published in 1942, would go on to become a key text in aquaculture and a driver in the coming shrimp revolution. The disaster of the war for the Japanese only confirmed Fujinaga's antinationalist tendencies.

After the war Fujinaga began carefully assembling a team of foreign graduate students so that the ideas he was propagating might go global. His ability to reach across nations stemmed from a natural warmth and good-heartedness. As one graduate student recalled, "He constantly challenged each of us to achieve but he also took time to sit down with his staff and talk about things other than work." He had a very simple dream, the graduate student recalled: "To make shrimp an affordable food."

But before shrimp could become an affordable food for the masses, Fujinaga and his successors had to help shrimp solve the three essentials of life: food, sex, and death.

Feeding was the first problem they tackled. In the wild, shrimp are opportunistic and omnivorous eaters, and traditional extensive shrimp farmers relied on natural organic matter to feed their shrimp. A closed system is very different. It took Fujinaga years to develop the right sequence of food to accommodate the three distinct life phases of juvenile shrimp. Along with diatoms, eggs of the Pacific oyster proved to be one of the first successful early feeds. From there, when shrimp passed from the mysis phase to the prelarval phase, Fujinaga found that a shift to Artemia shrimp (later popularized in America as sea monkeys) did the trick. And lastly, as they entered the postlarval phase, short-necked clam meat was the food of choice. This progression of feed was to become the key to unlocking the feeding not just of shrimp but later of other marine species. Indeed, shrimp was one of the creatures that opened the door to large-scale marine domestications and the rising aquaculture empires of Thailand, Taiwan, and China.

After Fujinaga finished training his foreign students,

most went back to their native lands to use his methodologies on their home shrimp species. From Taiwan to Thailand to Vietnam to the United States they set to work. With all these researchers in all these diverse places, they were able to share information on what worked and what didn't over a broad spectrum of environmental conditions. One of Fujinaga's heirs, I. Chiu Liao in Taiwan, found that *Litopenaeus monodon*, or the giant tiger prawn, grew the fastest and took best to a kind of high-density form of production that became known as "intensive" culture. As the Gulf shrimp boom surged in America in the 1970s and 1980s and chains like Red Lobster grew, major U.S. corporations realized that if they could figure out a way to have access to more and more supplies of shrimp they could have a very lucrative market. Increasingly, American researchers and businesses started reaching out to Asia, building shrimp collaborations on a number of different levels. Ralston Purina, formerly a company that produced food exclusively for land animals, began manufacturing a dedicated line of shrimp feed. Other companies and researchers tried to grow shrimp in Florida, Texas, and elsewhere in America and went so far as to lease Fujinaga's rearing technology. But they soon realized that labor costs would be much cheaper in South America and Asia. Over the course of twenty years, corporations like Du-Pont invested in shrimp aquaculture systems and research around the world. The tiger prawn began spreading along with that research as the most desired shrimp.

But the tiger prawn had one serious problem, a problem that Fujinaga did not have time to adequately address: sex. For some reason the tiger prawn had difficulty spawning in captivity. For an equally mysterious reason, it turned out that if you clipped off one of the tiger prawn's eyes it suddenly became fertile. This secret was unlocked in 1973 and soon after tiger prawn harvests soared. The secret and the larvae were traded back and forth between Taiwan and Central America, the world's two emerging shrimp-growing capitals.

The frenetic global swapping of shrimp from research station to research station and then from country to country eventually brought shrimp farmers up against their third and most formidable problem: early death. Shrimp are extremely primitive animals, having evolved hundreds of millions of years ago. As such, they don't have much in the way of an immune system and are notably susceptible to disease. Just as tiger shrimp started gaining global momentum, a bacterial disease called vibriosis appeared that leveled the industry. In an effort to mitigate its effects, farmers synthesized feed laced with the powerful antibiotics nitrofuran and chloramphenicol, among others—antibiotics that have since been linked to cancer in humans. In spite of the banning of these substances and efforts by Southeast Asian countries to adhere to Western food safety standards, antibiotics are still in use in shrimp farming, and only about 2 percent of imported shrimp are tested by the FDA. Even in that small sampling, violations occur and shipments are on occasion returned to their country of origin. Tiger shrimp were also affected by a disease called yellow head, which appeared in 1990 and which caused the shrimp's thorax to turn bright yellow before the animal floated to the surface of the pond and died.

The tiger prawn disaster, though, was not the end of shrimp. Quite the contrary: it led to the discovery of a new shrimp. As the authors and shrimp researchers Anne and Jack Rudloe write, the shrimp bearing the scientific species name *vannamei*, known more commonly as either whiteleg shrimp or Pacific white shrimp, hails from clear across the Pacific, thousands of miles away from the shrimp kingdoms of Southeast Asia. It had originally been discovered by a Dr. Vanname in 1926 in a market in Panama and lost in a pickling jar in the bowels of the American Museum of Natural History. When it was later described by a Miss Pearl Lee Boone, it became known as *Penaeus vannamei* Boone (and later reclassified as *Litopenaeus vannamei* Boone). Rediscovered in the late 1980s, it was soon repurposed toward aquaculture. By 1992 the whiteleg shrimp had spread around the world and now makes up more than 80 percent of all farmed shrimp production.

In the world of disease, however, nature always ups the ante. While *L. vannamei* were better at resisting bacterial infection than tiger prawns, they were prey to an even more deadly attack: a virus called "white spot." White spot had no cure and for which no known inoculation existed. Not heeding the lesson of bacterial infections in tiger prawns, farmers shipped whiteleg shrimp larvae from nation to nation from Asia to Latin America to Africa. White spot traveled with it. Soon the dreaded disease was on the move. Within a few months white spot had spread around the globe. Entire farms were wiped out in a matter of days. Anne and Jack Rudloe report that in the massive shrimp-growing region near Qingdao, north of Shanghai, the atmosphere above the rotting ponds was so fetid that passengers aboard airplanes flying over the region were hit by wafts of putrid air, causing them all to be engulfed in a wave of nausea.

But the most pernicious thing about shrimp culture and the diseases it propagated was the spurring of the phenomenon of pond abandonment. Ponds afflicted with bacterial infections were treated with antibiotics that would accumulate along with other wastes on the murky bottoms. Eventually it became more efficient to move on to dig new ponds than to clean up the old ones. This led to a daisy-chain effect. Mangrove forests were cleared, new ponds were created, waste accumulated, diseases appeared, ponds were abandoned, and then, in turn, new mangrove forests were cleared. High irrigation and energy demands also contributed to intensive culture's damaging effects, draining freshwater from natural aquifers. Digging of shrimp farm channels allowed salt water to penetrate inland in Vietnam and Bangladesh. Lands that were being inundated with salt water had formerly been good land for a historically much more important subsistence crop: rice. Shrimp, the miracle protein that promised to alleviate malnutrition, was in some cases exacerbating it by undermining the cornerstone carbohydrate of the vegetarian diet of coastal residents.

I witnessed the repercussions of pond abandonment as I flew above Ca Mau, Vietnam. While the amount of mangrove forest lost to shrimp farming is generally cited at around 5 percent globally, in Vietnam the losses are much more severe. From 1969 to 1990 the country's million acres of mangroves shrank by 33 percent. A more recent report states that the mangrove area in Vietnam in 2008, the year of my visit, was down to 148,000 acres—a decline of 86 percent from the 1969 area. It is no small irony that the first major wave of deforestation happened in the late sixties and early seventies as a result of American planes spraying mangrove canopies with the defoliant Agent Orange. The second wave of deforestation from the 1970s to the present is in part the result of American consumers eating shrimp.

With the mangroves cut away, storms can more easily wash over the coast, degrading marshes throughout, leaving this already watery part of the world open for catastrophic damage as climate change gets ready to raise sea levels everywhere. Bangladesh, a low-lying estuarine country the world knows best because of the flooding that displaces millions of people every year, is also a leading country in the production and export of farmed shrimp—mostly to the United States and Western Europe.

Yet there are other, much better ways to go about farming shrimp that take the living estuary into account and make it part of a sustainable equation. Which was why I had traveled to Vietnam and ventured up the Ganh Hao and then made a dogleg into the Bay Hap River and then finally found my way into a still wild mangrove thicket known simply as Farm 184.

Tying up our motorboat at the dock, my guide, Suong, told me to make sure that I held on tightly to my knapsack.

The forest was packed with monkeys, and with frequent foreign visitors to the farm, they had grown adept at stealing luggage. Monkeys were not the only wildlife; snakes, crocodiles, and all the various dangerous and not altogether appealing life of the wild swamplands were doing quite well on the protected grounds of Farm 184. They were doing well because Farm 184 represented a different vision: a compromise between some of the technology that had been introduced during the raging shrimp gold rush of the twentieth century and the older, less intensive traditional systems that had once been the norm. A synthesis that respected the natural boundaries of shrimp density, prohibited the practice of pond abandonment, and yet made full use of the best advancements in shrimp breeding and trait selection.

After the white spot epidemic of the 1990s, shrimp producers realized they would need to reform their methods. Looking at the advances other fish-farming sectors had made through selective breeding, particularly the salmon industry, shrimp producers in Thailand began breeding *P. vannamei* shrimp and arrived at a genetic line that is resistant to white spot disease. They figured out ways to aerate their ponds and deal with the flocculent ooze that accumulated on pond bottoms. They arrived at what are called "biofloc systems"—means of biologically reprocessing wastes to limit and even eliminate the discharge of polluted water from ponds. They developed feeds that are derived less from wild-caught fish—feeds that are less apt to rot and provide a breeding ground for disease. Finally, many producers began adhering to international production standards that ensure a code of behavior. Antibiotic use has declined markedly. Labor standards have improved. Even mangrove deforestation is starting to be addressed. Vietnam is in fact the leader in this regard. Notable efforts have been made to try to reforest lost portions of mangrove, albeit with mangrove mono-crop-style plantations that lack the species diversity of the original forests. Currently the country plans to plant an additional fifty thousand acres of mangrove by 2015.

But all of these plans to reforest the country's lowlands may run into the roadblock of shrimp farming. The basic premise, the idea that tens of thousands of acres of precious coastal wetlands should be given over to shrimp monoculture, has not been challenged. Some even argue that since Southeast Asia is so consistently—indeed, oppressively—hot, it should be the de facto homeland of the worldwide shrimp industry. "It's the *Guns, Germs, and Steel* argument," Josh Goldman, an American aquaculture entrepreneur who farms barramundi in the United States and Vietnam, told me. "My hypothesis is that tropical species and environments have a natural advantage. Since fish and shrimp are cold-blooded, their growth rate and food conversion are more efficient and stable where temperatures are hotter and less subject to seasonal variations; and that favors the tropics." Whereas the Gulf of Mexico produces one crop of wild shrimp per year, hot Southeast Asian farmed-shrimp ponds churn out two, sometimes even three crops of shrimp per year.

But while intensive shrimp farming has indeed made it possible for large amounts of shrimp to be produced—as much as eighteen thousand pounds per acre of pond every three months, or thirty times as much protein as is turned out by an acre of cow grazing land—very little of that protein has found its way into the mouths of the Vietnamese. Shrimp has become one of Vietnam's main cash crops. Areas that had been traditional subsistence fisheries for local fishermen have been made the property of shrimp companies. Just as sugarcane, indigo, and tobacco in colonial times displaced actual food and turned indigenous food systems into commodity trading systems, in the postcolonial era shrimp has converted what was once uniquely subsistence land—inhospitable marshy waterfront—into another Western-style cash crop system. These areas no longer produce food for local people to live upon. They produce a cash crop for foreign consumption—in Vietnam over a hundred million pounds of shrimp a year—enough for every Vietnamese to eat more than a pound of it annually if they could get their hands on it. But they can't: over 70 percent of all Vietnamese shrimp goes abroad, 90 million pounds of it to the United States alone.

Which was where Farm 184 represented something different. Instead of practicing intensive shrimp culture, Farm 184 was taking a gentler approach. Yes, it was taking full advantage of the products of the revolution Fujinaga had started. It was using selectively bred, fast-growing shrimp that were the product of fifty years of genetic selection. But the farm was using a system of production that was far friendlier to its home environment than the shrimp production systems the world market demanded. Farm 184 did use fertilizer to promote aquatic vegetation that fed its shrimp. And while it did sculpt the environment to allow water exchange between its ponds and the outside river, the environment was a good deal more wild in its function than intensive shrimp operations. Indeed, as I walked along Farm 184's elevated walkways, monkeys and chattering birds drowned out the roar of motorboats and it was hard to say where the farm ended and the forest began. It's important to note that Farm 184 is a standout even among other more sensitive farms. It is state owned, existing within the confines of a nationally protected mangrove reserve, and has received technical assistance from the Swiss food company Sippo. Sippo, like many European producers, is increasingly active in Southeast Asia in trying to obtain lines of safer and at times organic product and will contribute to their research and development.

Of course, having a farm with a high degree of environmental protection has its economic costs. A farm like 184 grows only a fraction of the shrimp an intensive operation grows. But the lower productivity did not affect my host's generosity. As I left Farm 184, the old motorboat we'd come in on was loaded up with tiger prawns and whiteleg shrimp and before long we were flying back upriver past small floating restaurants, past children plunging into the brown water, past temples of a half dozen faiths until finally we arrived at Restaurant 257 in downtown Ca Mau.

The cooked shrimp were brought out to us shortly thereafter, most of them whole, lightly poached in coconut milk. To say that they were eaten plain is a bit of a misrepresentation. Everything in Vietnam comes garnished with the spoils of the country's almost ridiculously bountiful agriculture. On the table were all the fixings for do-it-yourself summer rolls: a pile of elegantly latticed rice paper sheets, pineapple, green banana, shredded banana flower, and four or five varieties of herbs and greens I couldn't really place. There were other incarnations of shrimp—shrimp sautéed with sweet lotus root and braised in a coarsely ground black pepper sauce. Other items on the menu included *thien ly xao toi*—yellow flowers sautéed in garlic—and alien-looking *ca keo*—goby fish baked in a clay pot with sugar and *nuoc mam* sauce. Sweet and tangy white-fleshed mangosteens rounded out the meal.

The quality of all this fresh shrimp was notable. Had they been frozen and processed by an intensive shrimp farmer, it's likely they would have been treated with tripolyphosphates—a common chemical that increases the weight of seafood before it gets to the consumer. Phosphates occur naturally in muscle fiber, helping muscles retain water and thus lubricating and cooling them during strenuous activity. Food scientists in the 1950s realized that they also can work to keep frozen foods succulent, a fact that proved to be a temptation to many shrimp producers. When seafood retains water, it comes out heavier on a scale; a heavier shrimp brings in more money. Unfortunately, the overuse of tripolyphosphates has become commonplace in Asia. Though not a health risk per se, a shrimp loaded with tripolyphosphates is apt to shrink in the pan. Shrimp treated with phosphates remain eerily translucent even when cooked and can have an overly chewy consistency. It's for this reason that, in the American South, Asian-farmed shrimp are sometimes called "gummy bear shrimp."

No, these Vietnamese shrimp were most definitely not

gummy bear shrimp. But I couldn't help but think that this grand shrimp banquet was too much for me. By all rights, it should not have been mine at all. Though I liked the way these shrimp had been produced, liked how their production was integrated into the life of the mangrove forest and took very little from nature, the company that was selling these shrimp didn't seem to look at it this way. Their perception seemed to be that some Westerners like organic shrimp and some like regular shrimp. Shrimp for these producers were like so many widgets of varying quality, some of which fetched a higher price for an overly fussy elite while others drew standard prices for less well-heeled customers. Embracing in a rudimentary way the precepts of Western "choice," Vietnamese growers offered both options and seemed to shrug their shoulders as to the reasons why.

And therein lay the problem. Having this choice in the first place was not really what Vietnam and indeed all of Asia needed. What coastal Southeast Asia needed was good, safe, nutritious shrimp and stabilized shorelines—exactly the kind of thing that could be offered by converting the country to the kinds of methods practiced at Farm 184. By that reasoning, shouldn't these well-farmed shrimp have been produced for the people who lived here? There are about eighty-five million people living in Vietnam at present. If all Vietnamese shrimp farms were converted from damaging intensive shrimp culture to improved extensive shrimp operations, they would produce much less shrimp. Not enough to maintain the major export market the country enjoys, but just about enough for the average Vietnamese person to have some shrimp from time to time.

Shouldn't every last bite of very good Vietnamese shrimp belong to the Vietnamese? Shouldn't every last mile of Vietnamese coastline be protected by nourishing mangrove forests so that this populous and growing nation doesn't lose any more land to the incursion of the sea, an incursion that is surely to grow as sea levels rise over the next century?

None of these questions really even occurs to the giant U.S. wholesalers who make use of all this Asian product. To the contrary, once American wholesalers became addicted to farmed Asian shrimp, they only wanted more. Our chains and warehouse stores require a constant supply of uniform product that can be "value added" and branded with flavorings and processing. This is due to a process that World Bank aquaculture analyst Dr. James Anderson and Dr. Quentin Fong noted in a recent paper. According to Anderson and Fong, there are several characteristics necessary for a food product to transition from a local food to a global commodity:

1. product homogeneity

2. storability and year-round supply

3. transparent and efficient cash pricing

4. the ability for traders and producers to forward contract

Asian shrimp provide all of this.

Homogeneity was achieved with the universal acceptance of whiteleg shrimp as the industry standard. Though there are more than two thousand species of shrimp in the world, humans have zeroed in on this single species as the one they want to eat.

Storability and year-round supply was achieved by the perfection of freezing techniques and the multiple crop yields that intensive Asian shrimp farms were now yielding.

"Efficient" pricing was achieved by reduced labor costs possible only in nations with low wage standards.

And finally, the ability for traders and producers to forward contract—in other words to predict how much shrimp people would eat—was provided by Americans themselves. Fueled by an ever declining price and an ever increasing appreciation of shrimp's neutral taste and mouthfeel, traders and producers could be more or less assured a steady demand in shrimp coming from the United States. Shrimp became so predictable that in 1993 the Minneapolis Grain Exchange, an important hub for trading commodity products, established a shrimp futures market. People on Wall Street who had no idea where shrimp came from or the relationship they had with the world's ecology could trade the future sales prices of shrimp in the same way they were now trading beef, oil, and gold.

Once shrimp had been turned into an international commodity, everyone wanted in, even American fish processors that technically should have been in a trade war with Asian producers. Increasingly, American fish houses—the coastal processors that had historically dealt in American-caught fish—found that they could buy frozen shrimp from Asia and repackage it for less than what it cost to buy and process American shrimp. They did so without any qualms, further devaluing the wild American product and making the supply chain for Asian and American shrimp blurrier.

Not only has foreign aquaculture put extreme price pressure on the American wild fish we eat, it also directly competes for American resources. China alone imports seventy-five million pounds of fish meal derived from American wild fish, which it uses to feed its aquaculture industry. A large portion of that fish meal comes from a fish called a menhaden—a keystone prey species to wild food fish throughout the Gulf coast and the Atlantic Seaboard.

E ventually Asian producers augmented their product lines, diversifying the species they started sending to America and the rest of the Western world. In the 1990s Vietnamese Pangasius catfish, a fish with a very different taxonomy but the same mouthfeel as American catfish, slipped in and is now the sixth most consumed fish in America. When a consortium of American Southern states pushed for tariffs against that fish and mandated that it not be called catfish but rather by the Vietnamese names *tra* or *basa* or *swai*, China began growing traditional American-born species of catfish on Chinese farms and then selling them back to the United States at significantly lower prices than that of American-raised catfish. In 2006 and 2007 the U.S. Food and Drug Administration restricted Chinese exports of catfish to the United States after finding traces of carcinogenic coloring and antibiotics. The stricter rules have made China better enforce regulations, but the sheer number of producers in China makes it difficult to check every farm thoroughly.

And there have been other notable examples of American species taking up residence in Asian aquaculture only to be sent back to us for our consumption. In 1980 Professor Fusui Zhang began discussions with American aquaculturists about the possibility of bringing the United States and China into closer collaboration. Choosing the fast-growing New England bay scallop as a possible candidate, Zhang tried several times to bring over scallops that had been grown in U.S. laboratories (either the Milford Laboratory of Loosanoff fame or the Virginia Institute of Marine Technology—the historical record is not clear). Finally, in 1982, Zhang traveled to Martha's Vineyard and selected two hundred scallops from the Martha's Vineyard Shellfish Group on Lagoon Pond in Tisbury. Of the two hundred that made the journey to China only twenty-six survived. But they would go on to spawn an industry that is now valued at half a billion dollars, with over a thousand hatcheries. Those very cheap five-pound bags of scallops that the bargain hunter may on occasion find in the supermarket frozen-seafood bin are the direct descendants of the Zhang-created scallop diaspora. Examples of this kind of intentional species relocation are in part a result of our country's uneasy relationship to aquaculture. On the one hand, some of our finest government-funded universities and research institutions produce numerous scientists adept at helping farmers develop and grow farmed seafood. On the other hand, our country's regulations and coastal zoning restrictions mean that there are very few places in this country where seafood can actually be farmed. As a result, many of these excellent American scientists fail to find work in their home country and look elsewhere.

And so, more and more, our seafood comes to us from foreign aquaculture. Instead of wild New England cod, fish and chips are often Chinese-farmed tilapia. Instead of wild striped bass grilled whole and set on a plate, American size restrictions mean that smaller, plate-sized fish must come in the form of Turkish-farmed European sea bass, aka branzino aka loup de mer. Even redfish, that esteemed Southern fish that the New Orleans chef Paul Prudhomme famously drove into decline when he created "blackened redfish," now rarely comes from the Gulf of Mexico. A redfish on an American plate is much more likely to come from a farm in China than a port in the American South.

And what does all this do for the American marsh that supports all the wild American seafood? Nothing good.

The Shrimp Homeland Sinks

"Did you hear . . . alarms on the evening of April 20?"

"No, sir."

"Do you know why you never got that alarm, sir?"

"Yes. They were bypassed."

"And how do you know they were bypassed, sir?"

"Because I physically seen it on the screen. They were actually what's considered 'inhibited,' not bypassed. The correct word is 'inhibited.'"

Thus went an exchange between an investigator and the chief electronics technician aboard the Deepwater Horizon as they reviewed the events of April 20, 2010, the night the rig failed and blew and began an ordeal that spilled millions of barrels of oil into the Gulf of Mexico. The recorded testimony was being reviewed in the federal court building on Poydras Street in New Orleans and marked the beginning of the latest round of a process that had been grinding on for three years. BP had once again raised its defenses and was making a firm stand against what could be a penalty worth tens of billions of dollars under the Clean Water Act.

The company had recently balked at a $16 billion settlement agreement with the Environmental Protection Agency, claiming that the state of Louisiana was now demanding more money than had been initially agreed upon. And now in the spring of 2013, in the grand, balconied federal court building, thousands of dollars were pouring through the floorboards as twenty-odd lawyers and other suited professionals listened to this excruciatingly detailed testimony on

Horizon's "integrated automatic control system" and how it may have allegedly been put in "inhibited" mode.

This felt surreal given the much more urgent alarm system blaring from the great shrimp homeland of southern Louisiana. Thanks to a half century of bad behavior by oil companies, the land where the Poydras Street courthouse sits is in danger of disappearing altogether. The state of Louisiana is currently losing marshland at a rate of a football field every hour, land that was once critical to rearing shrimp and protecting New Orleans. That those of us outside of shrimp country haven't noticed is due to the fact that we, like the Deepwater Horizon's alarm system, have been put in inhibited mode ourselves. We hear no blaring warning that we are losing our shrimp because our markets are awash in farmed foreign product that masks the sound of the alarm.

That alarm started blaring with the discovery of oil in Louisiana in 1924. From the outset of petroleum exploitation in and around the delta, developers began dismantling the elements that underpin the great marsh's stability. As Rowan Jacobsen relates in his fine book *Shadows on the Gulf,* incursion into the marshes began when oil companies dredged shipping channels to allow shortcut access to the refineries along the coast and upriver. Eventually channels were expanded into a spiderweb of interconnecting routes.

Channels were in turn paired with a pipeline network laid underground. Natural winding shallow channels were replaced by straight, deep canals; levees of dredged materials were piled up in unbroken lines parallel to the canals. These "spoil banks" interrupted what's known as "sheet flow" across the wetlands—the regular refreshing and cleansing of the marsh that takes place during floods. In the absence of these revitalizing sheet flows, organic material accumulated and rotted. In addition, cuts into the marsh allowed salt water to flood into areas that had typically been less saline. Marsh grasses that are accustomed to lower salinity died. Without live roots holding the mud together, things started to disintegrate. Now the Louisiana marsh, that great shrimp nursery at the end of the river, is falling apart.

The erosion of the Louisiana marshes is being further accelerated by another incursion: the powerful agribusiness of the Mississippi valley. The Mississippi River that feeds into Louisiana is one of the most engineered rivers on the planet. The reason it was engineered to such an extent was so that landfood producers could gain access to the tremendous fertility the river deposited in its meandering floodplain.

As John Barry relates in his Mississippi River history *Rising Tide*, James Buchanan Eads was one of the first to figure out how to control the river and claim the rich floodplains. In 1833, Eads arrived penniless in the frontier town of Saint Louis at the age of thirteen; he soon launched a lucrative salvage business, rescuing everything from steamboats to blocks of lead from the river. Inventing a primitive diving bell, Eads walked the river bottom. He intimately "felt" the river and understood the nature of its power. Eventually he would imagine a river-long system of levees and diversions that would start in Cairo, Illinois, and extend to the eponymously named Port Eads, Louisiana, pinching the river and draining the floodplain.

Eads's most devastating idea was even more radical: to turn the river from a lazy, wandering affair into a straight line. Although this did not occur in Eads's lifetime, his proposal was taken up after the record flood of 1927.

Throughout the 1930s and 1940s, the Army Corps of Engineers systematically lopped off what had been called the Greenville Bends and a dozen other large meanders, shortening the river by 150 miles. Floods were indeed reduced, but afterward the lower Mississippi transitioned from being a complex marshy wetland into a fire hose that blasted sediment straight into the Gulf. With the Mississippi River radically altered, straightened, and no longer able to deposit new marshland in the delta, shrimp habitat loss was made even more extreme. In addition, nitrogen-based synthetic fertilizers coming from the corn-growing heartland surrounding the Mississippi were also shooting out into the Gulf. Formerly, when the river was more curvaceous and the floodplain was a hundred miles wider than it currently is, fertilizers and silt were able to drop out of the river and spread out over the entirety of the Mississippi valley. The floodplain performed a kind of dialysis for cleansing the water of nutrients. Today all that fertilizer goes directly into the Gulf, causing extensive algal blooms. And when those algae die, they are consumed by bacteria, which in turn suck life-giving oxygen from the water.

This problem is further exacerbated by big agriculture's devastation of what had previously been one of the world's largest tracts of an ecosystem called "bottomland forest"—another critical biological filter. "If you look at it historically," Keith Ouchley of the Nature Conservancy of Louisiana told me as we stood in a remnant forest buzzing with cicadas and laden with vines and green growth of every kind, "from the southern tip of Illinois to the Gulf of Mexico there were twenty-four million acres that were nearly 100 percent in forested land. It was the largest temperate deciduous floodplain forest in the world." Much of this was intact as recently as the 1950s. But then came soy. "In the 1950s soybeans came into vogue," Ouchley explained. "Those low, wet areas that had previously been too tough to farm began getting cleared. I remember as a kid the bulldozers coming into tracts of woods and taking down all the trees. They'd roll 'em up and burn 'em. They wouldn't even harvest the timber. There were columns of smoke all over the landscape. It was like the slash-and-burn things they do in the Amazon."

Thanks to the removal of all of those critical biological filters and the addition of so much chemical fertilizer, the outflow of the Mississippi is now markedly more nutrient rich. So much so that massive algal blooms occur every spring and summer around the river's outflow. When those algae die and decompose, oxygen is removed from Gulf waters. And now, every year, a swath of water as large as the state of New Jersey forms in the Gulf that is so low in oxygen that creatures like shrimp must flee to other waters. This so-called dead zone has been forming annually in summer months at least since the 1970s. Shrimpers are now driven farther and farther offshore in search of a decent catch. In effect, we are trading seafood for landfood, favoring industrial agriculture over a productive natural food system.

So why, then, the boom of shrimping in the 1970s, 1980s, and 1990s? With wild fisheries there are always multiple processes at work, and there are of course many different cyclical causes of booms and busts. But the most disturbing possible cause for the shrimp boom is one that Randy Pausina, the assistant secretary of the Louisiana

Department of Wildlife and Fisheries, described to me. "Shrimp like to feed along the fresh edge of a coastline. The more edge there is, the more nutrients there are, and therefore the more shrimp there are. When the coastline deteriorates, it gets jagged and creates more edge—more shrimp habitat. So as the marsh retreats you get a false sense of boom in production. The problem is we don't know when it's all going to crash. Is it going to be next year? Ten years? A hundred years? We don't know." But when the Gulf shrimp population does finally begin to seriously decline, we are not likely to notice. Because the foreign shrimp-farming industry grew exponentially during the last thirty years, it has all but eclipsed the presence and identity of American shrimp in the marketplace. If we finally lose our own shrimp, who will even notice?

For nearly ninety years, as the oil industry canalized the marshes and commodity agriculture turned the Mississippi into a fertilizer fire hose, shrimp habitat has been fading. Now, as teams of lawyers and scientists try to assess the damage of the 2010 spill, it is extremely difficult to differentiate between the damage the Deepwater Horizon caused by releasing nearly five million barrels of oil in three months and the damage big oil and big ag did by destroying thousands of acres of marshland and bottomland forest over the course of a century. Both are most certainly adding to the marsh's deterioration. As Garret Graves, chair of the Coastal Protection and Restoration Authority of Louisiana, told me, the spill accelerates what was already a troubling process. "The marshlands have a protective edge of grasses that hold the shore in place. When that first wave of oil hit the coast, it killed that leading edge of grass. When that dies, it will fall away and we're going to lose even more land." And this is no small amount of edge affected. In all, according to the Louisiana Coastal Protection and Restoration Authority (CPRA), more than six hundred miles of Louisiana coastline were oiled by the Deepwater Horizon blowout. Given what we know about the effect of oil on coastline, it is reasonable to assume a retreat of marsh along those six hundred miles.

Indeed, the food effects have already been felt in the Gulf. Reminiscent of an earlier time and another place—namely, New York Harbor in the early 1900s—it is the Eastern oyster that has sounded the most immediate alarm. The Louisiana oyster grounds are the largest extant wild oyster beds in the United States and perhaps in the world. They are the advance edge of the marsh, the keystone species that holds the whole thing together. And they are suffering. "On the first Wednesday following Labor Day the oyster season in Louisiana has opened for public oyster harvest for over a hundred years," Al Sunseri, president of the

136-year-old P & J Oyster Company—the longest continually running oyster company in America—wrote me in the
fall of 2013. "The only problem with the 2013–14 oyster
season and the last three public oyster seasons following
Deepwater Horizon is that our oyster resources have been
remarkably nonexistent. You would think with all of the
continual BP commercials for the last three and a half years
that the Gulf of Mexico and the surrounding estuaries are
in pristine condition and everything is back to normal. I'm
here to tell you that is far from the case. Our fifth-genera-
tion oyster company stopped our shucking operation on
June 11, 2010, because of the lack of oyster production.
Soon we may have to lay off our few remaining employees
with my brother, my son, and I being forced to open the
shop only when we have a handful of oysters to sell."

But this needn't be the death knell of the Gulf marshes.
In November 2012, the federal government at last won its
criminal lawsuit against BP and the corporation agreed to
settle for a sum of $4.5 billion. This is just the beginning. In
addition, under the Clean Water Act, BP is potentially liable for a fine of $1,100 to $4,300 per barrel of oil spilled.
Indeed, this was the motivating factor behind BP's consistent contestation of the total number of barrels leaking from
the Macondo well. The more barrels documented, the
higher the total bill. And the bill could be very big—as

much as $20 billion, should the plaintiffs prevail in the next round of suits.

Thinking ahead to these possible resources, state planners in Louisiana have drawn up the "Comprehensive Master Plan for a Sustainable Coast," a fifty-year, $50 billion effort that seeks to stop Louisiana from sinking into the sea and, by extension, to save Gulf shrimp. A major part of the plan is the granddaddy of all shrimp savers: the rebuilding of the Louisiana marsh, a multibillion-dollar reconstruction effort, one that aims to restore the bio-filtration that all that expansive marsh provided in the past—the macrogeographic equivalent of a kidney transplant.

A day after my visit to the BP trial, I went on a road trip to one of these transplant operations with a group of officials and contractors working with the state's CPRA. In the community of West Pointe à la Hache, I saw a cluster of enormous white tubes rising up over the Mississippi's levee and extending into the sea. Hopping onto a boat, we followed this expensive piece of plumbing called a "sediment diversion" to its end. There, at the outflow, sat some of the newest land on earth, land that state officials hope will save not just shrimp country but all the densely settled human territory upstream. Given that sea levels may rise considerably before the end of the century, the project can have the feel of expensive sand castle building. And some ecologists, like Louisiana State University's Eugene Turner, believe that some of this artificial-marsh building can actually be detrimental to the long-term viability of the delta. What is really needed, Turner and others argue, is a fundamental rethinking of the industrial agriculture that is impacting Louisiana's wetlands.

Still, the vision of marshland being created before my eyes, with all its potential for nursing shrimp, was powerful. After West Pointe à la Hache, the CPRA van took me to a second reconstruction project a few miles up the river, near the town of Myrtle Grove. This older human-made mudflat was home to grasses and even some low-lying bushes. It looked like marshland. It felt like marshland. Could Louisiana figure out a way to make BP pay for more of this?

And though $20 billion is still short of the estimated $60 billion that would be needed to restore the Louisiana marsh ecosystem to an acceptable state, that money, well applied, could do a lot of good. In July 2012, Congress passed the RESTORE Act, which would commit 80 percent of BP's penalty fines back to the Gulf states. Normally Clean Water Act penalties go into a general federal kitty and often don't get applied where they are needed most. This time, hopefully, that will not be the case.

But in order for those penalties to go where they are needed, we as Americans need to feel that need. We need to understand that the marshes of Louisiana are not just an idyll to observe egrets and alligators; they are a food system, one that provides a large portion of the catch in the continental United States. If we choose to, we can support the environment that is home to shrimp, redfish, bluefish, blue crabs, oysters, flounder, sea trout, and others. Yes, there is a small risk of contamination from eating wild seafood from the Gulf. But that risk, when compared with all the other food risks we take as a nation, is infinitesimal. Daily we risk our health through salmonella in our eggs, *E. coli* in our beef, diabetes in the amount of high-fructose corn syrup we consume. Daily we risk contamination from poorly inspected farmed foreign seafood, less than 2 percent of which receives even a glance. All of these industrial food products we eat contribute to a degradation of the environment. If we could, somehow, fix shrimping and the other kinds of seafood extraction we practice in the Gulf, ultimately that relationship would be a good thing. We could perhaps nurture a class of fishermen who, like the shrimper Cal Aubrey, feel a personal responsibility for the seafood they bring in from their local waters. "This is my world here," Aubrey told me that night out on his shrimp boat on Lake Pontchartrain. "This is my livelihood. I coexist with nature. I don't destroy my environment. I make a living off of it."

How then are we to fix our shrimp problem? Is there a

way to reverse the tide, and move toward a different kind of model, one where the fisherman and the fish eater are more directly linked? One where that interrelatedness builds a sense of community rather than a global commodity?

This very question occurred to Thomas Hymel, a marine extension agent with Louisiana State University AgCenter and Louisiana Sea Grant based in Delcambre, Louisiana, a sleepy old bayou town in the heart of Cajun country, a town ruled by shrimp processors.

Prior to becoming a local fisheries marketing advocate, Hymel worked for Louisiana State University helping to educate shrimpers about the slew of regulations that came down in the 1990s—the requirements for turtle excluding devices, the seasonal closures, the bycatch reduction devices. But after the rise of Asian shrimp, after Hurricanes Rita and Ike, Hymel realized the very constituency he was supposed to be helping was in danger of disappearing entirely.

"What we've seen is that our fishermen have been put under so much pressure that a lot of them have just gone out of business. They are gone. It used to be that the boats in our harbor would get passed down to children. But there has been an exodus from business." Meanwhile, the people in the surrounding towns, be they the villages around Vermilion Bay or the Cajun capital of Lafayette, were not even eating local shrimp. Foreign shrimp were much cheaper. The shrimp they ate could just as easily have come from Ca Mau, Vietnam, as they could have from Delcambre, Louisiana.

What Hymel and others realized was that some sort of alternative distribution system had to be created—one that reconnected the consumer to the fisherman and at the same time elevated the price of shrimp out of the bargain basement of the commodity market. After receiving an initial grant of $5,000 from the local Twin Parish Port Commission, which oversees the Port of Delcambre, Hymel and his colleagues built a Web site called DelcambreDirectSeafood .com. Casting around for shrimpers who would join, they found only one who was game: an old grizzled fisherman named Gimmie Dupré. The idea was simple: Dupré would go out shrimping. He would post on the Web site from his boat what he had caught and then, hopefully, consumers would come to the dock and buy his catch. Formerly this would have been impossible. Shrimp processors had exerted such a monopoly on the Louisiana shrimp fishery that they had made it illegal for a boat to sell shrimp to consumers straight from the dock. But now with the shrimp processors in decline, no one stood to defend those old laws. Dupré was suddenly able to sell his shrimp to the people who wanted it most: Louisiana Cajuns.

DelcambreDirectSeafood.com was launched two weeks after the BP oil spill. After catching a load of white shrimp,

Dupré posted to the Web site and motored into port. He was hit immediately with a local shrimp flash mob.

From there, word of mouth spread from fisherman to fisherman. Now twenty different boats post to the DelcambreDirectSeafood.com site every day during fishing season. The flash mobs grow bigger and bigger. The shrimpers receive a price 50 percent higher than the processors pay, a full $1.60 more per pound. This translates into earnings in excess of $80,000 per boat per year. When I asked Hymel how many boats in all were in the program now, he didn't quite know. "But I'll tell you this," he said. "You can tell the boats in the program because they are the ones that have a fresh coat of paint." When it comes to bulk-buy frozen shrimp, small Gulf shrimpers can't compete with imported, farmed shrimp, but in this specialized market they are valued as part of a community and part of an environment.

Yes, these shrimpers, even with their better dollar-per-pound prices, are having the issue of bycatch. Even with the best available technology, shrimping still results in a bycatch rate of anywhere from two to four pounds of discarded fish for every pound of shrimp brought to market. But what shrimpers found was that once they had a chance to talk directly with consumers and explain what bounty existed in Gulf waters, they could begin to persuade them to try seafood beyond shrimp, to make use of all that wasted fish that's normally dumped overboard. Flounder, black drum, and blue crab that get caught in the shrimpers' nets are all snatched up by a population that is rediscovering its relationship with its own coast.

"We are reconnecting fishermen with their community," Hymel told me. "It used to be, years back you knew fishermen. But that all went away. Now people are plugging back in to all that again. They like these fishermen. They have great stories to tell. And it's refreshing to meet these folks if you've been working in an office all day. They are crusty old guys. They are not politically correct, but they love what they do and they do it in their style."

From the success of DelcambreDirect, Hymel realized that the relocalization of seafood was something that could work everywhere. Well, at least everywhere in Louisiana. And the Gulf States Marine Fisheries Commission thought so, too, and gave LSU and Hymel a grant to create a pioneer statewide direct marketing and education program called LouisianaDirectSeafood.com. Under this umbrella, three more geographically distinct, direct, indigenous programs were started, SouthShoreDirectSeafood .com, LaTerDirectSeafood.com (in Lafourche-Terrebonne), and CameronDirectSeafood.com, each boasting its own host of local fishermen. Each has even started to suggest that its particular shrimp might have a specific taste, a *merroir*, if you will. In the spring of 2013 the local shrimp industry in Delcambre, in partnership with Twin Parish and LSU, has moved forward with its own local brand of shrimp, named "Vermilion Bay Sweet" after the bay where its shrimp are harvested.

Initially Hymel received considerable pushback from the shrimp processors and traders on the docks—a treatment Hymel calls the "Bayou Blackball." Dean Blanchard, the shrimp broker I'd met in Grand Isle who had so humorously rebuffed the state's effort to retrain him for other work, was similarly aggressive with Hymel when he caught wind of what he was up to.

"Mr. Hymel," Hymel remembers Blanchard saying in an angry phone call, "are you trying to put me outta bidness?" Blanchard, from his side, argues that the DelcambreDirect model unfairly skews the game by allowing some shrimpers to in effect get a marketing subsidy. "They're using my tax dollars to pay a guy like Hymel to put me out of business," Blanchard told me. "The government is spending millions of dollars on folks that sell less than 1 percent of the product. They need to help the 99 percent." Blanchard points out that three years after the BP blowout three key shrimping

grounds—Bay Jimmie, Grand Terre, and Elmers Island—
all remain closed to fishing due to oil spill concerns.

But the ones who are catching shrimp are finding good
prices for a change. Not only are lower supplies from the
Gulf raising the price. More significantly, bad news for
shrimp farming is coming out of Asia. In 2009 a brand new
shrimp bacterial disease called early mortality syndrome, or
EMS, appeared in China. The same trading of shrimp larvae from facility to facility that had spread vibriosis and
white spot has done the same for EMS. Within a few years
the disease had migrated to Thailand and Vietnam. In 2013
the disease caused losses to the Asian shrimp market of over
$1 billion.

So, times just might get good again for the small-time
shrimper in the Gulf of Mexico. For a while anyway. But
even Thomas Hymel, the architect of Gulf shrimp direct
marketing, is leery about the future. As one immersed in
the fishing industry, he too worries about the marshlands,
that indispensable yet ever so fragile heart of the fishery.
"The sustainability of our fishery is all linked to the health
of our marsh," Hymel told me. "There's gonna come a point
when we don't have what we have. If the marshes disappear,
our fisheries will disappear. That's just clear to everybody
who plays in this arena. We've been steady in our shrimp for

the last few years. But we all wonder where that tipping point is. At what point do we see this thing go in a different direction?"

At what point indeed? Perhaps that point will come only when the environmental value of local seafood is recognized not only in the bayous of Vermilion Parish but in the remaining mangrove forests of Ca Mau, in the vanishing wetlands of China's rapidly industrializing Pearl River delta, and in the slowly recovering former shrimp grounds of San Francisco Bay. We have a choice ahead in this regard. We can continue down the path of the last half century, treating the fewer and fewer seafood species we identify as food like so many widgets, domesticating them, farming them, and turning them into units of protein that fit between the two halves of a bun—"dough delivery systems," as one fisherman wryly remarked to me on the current status of seafood in our society.

Or perhaps we could take a pause and reacquaint ourselves with the specificity of the fish and shellfish on our plates. We could come to understand that shrimp are part of a web of life integrally connected to the health of our immediate environment. And in the process we might learn that this environment is as fragile as it is necessary, that it requires our ecological stewardship as much as it does our economic interest.

Sockeye Salmon

Amid the bleak attitudes that pervade American reporting on fisheries today, it is easy to think of the fights around seafood as battles already fought and lost. The great oyster kingdom of the Northeast is a remnant of what it once was; all we can do is attempt to piece together a few artificial reefs and try our best to expand upon the oyster-farming industry that has supplanted wild systems. The Gulf of Mexico marshes that once churned out so much wild shrimp are slipping through our fingers, a mere 50 percent of what they once were—and we must pump in billions of dollars just to keep the other half from falling into the sea.

But dwelling too much on the tragic aspects of the American relationship with natural seafood systems may obscure an important fact: the United States still has a hell of a lot of fish.

In fact, there is one place in America where there is so much fish that we don't really quite know what to do with it all. The subtleties of this place's abundant wild fisheries are astounding. They contain millions of interconnecting biological threads hitched to levers we haven't even begun to identify. By obsessing over the ruined and half-ruined waters of our country, we run the risk of ignoring this last great place that produces 5.3 billion pounds of seafood a year—a full half billion pounds *more* seafood than Americans consume every year.

This last great place is a country called Alaska. Yes, it is technically just a state and has overlaying it a skein of American rules and order. But its statehood is only half a century old, and as the nations of Asia begin to exert the tug of their growth and proximity, Alaska has entered into a kind of twenty-first-century Great Game, behaving more and more like its own proper nation—a highly unusual nation with many big plans.

One plan on Alaska's books is to construct an eight-hundred-mile pipeline underneath the entire north-to-south center of the state that will serve as a conduit for shipping Alaska's abundant natural gas to China, Korea, and Japan. Alaska tried to do this kind of thing with another nation, one to the south. But that other country, the United States, decided that building a two-thousand-mile pipeline under mountain ranges and lakes might prove too expensive and potentially harmful to the territory the pipeline would traverse. Being opportunistic and seeing Asia's natural gas consumption rise, Alaska went out and sought other partners.

Another plan Alaska has is to drill into its aquifers and figure out a way to sell its water abroad. China has polluted its public waterways to such an extent that it is in dire need of clean freshwater. But even with China's clean-water deficit, one cynic in Alaska noted to me that some of that imported water is surely going to end up in plastic bottles and exported from China and sold back to us here in America as pure Alaskan spring water, bottled in China. Whatever its eventual destiny, Alaska has plenty of clean freshwater for the moment and is happy to oblige.

But one of the biggest plans Alaska is dreaming up concerns a bay located just over the Chigmit Mountains—a thin, high ridgeline that separates the rapidly changing lower third of Alaska from its still mostly wild upper two-thirds. It is the bay from which former governor Sarah Palin took a name for her daughter, though the personal connection has not warmed her heart. In fact, Palin appears to support serious industrial exploitation of the bay.

Some say that when you look at a map of the state of Alaska, the state looks like a human head in profile facing west toward Asia. In such a profile, this bay is located roughly where the mouth would be. If we rip out this mouth, it will be at our peril. Because this bay contains one of the greatest seafood treasures of all: the biggest sockeye salmon run left in the world. It is a run that generates as much as two hundred million pounds a year of some of the purest, most nutritionally rich wild protein on the planet. It is a key to our nation's food security, health, intelligence, and endurance. If ever there was a place where the American catch is making a crucial stand, Bristol Bay, Alaska, is it.

Copper in the Spawning Grounds

"Okay, now here's the safety lecture," Rick Halford said as he looked over his shoulder from the pilot seat in his little Cessna 185 floatplane. During my trip to Bristol Bay the airplanes kept getting smaller and smaller as I passed from New York to Anchorage to Dillingham to a lodge on the Nuyakuk River until finally I was in Rick's plane with my knees squinched up and my elbows tucked in to avoid the control wheel. "The doors open like this," he said, wiggling the handles and opening the door over the river's roiling waters. "The seat belts are like this"—*clickety-click*—"and in the back there's a bag with some food and some other stuff you might need." End of safety lecture.

In spite of the bluntness of the safety information, Rick inspires confidence as a pilot. This is odd, because Rick is a man who mostly taught himself how to fly.

Rick comes from an old New England Yankee family and proudly asserts that, unlike most of the four generations of Halfords before him, he did *not* attend the elite Maine college of Bowdoin. Back in the 1960s he decided to say good-bye to all that and drove to Alaska. Once across the border, he continued until he could drive no farther. And since what primarily interested Rick was what went on in Alaska *after* you couldn't drive anymore, he realized in short order that an automobile was the wrong sort of vehicle for his future life. He sold it for thirty-five hundred dollars and bought his first airplane. It took only a few hours of rudimentary instruction for him to begin flying solo, and he has been flying various planes ever since.

"This plane has hauled a lot of hunters and a lot of miners. It's never hurt me and I've never hurt it. And now we're friends," he said as he fiddled with all sorts of levers and handles that to my flightless self seemed not at all friendly. And then, whipping the plane around into the wind and zooming down the river, he made a little jerk on the throttle with about as much thought as throwing a car into drive and suddenly he engaged that other gear that puts your vehicle not forward, not backward, but upward into the sky. He then one-handed the yoke as we settled into a low-flying cruise, breaking northeast toward the mineral strike called Pebble Mine.

For many years Rick Halford had been a very contented Republican. In the 1970s he started up a guiding business in which he happily flew hunters in from the lower forty-eight. This wasn't always the easiest of jobs. Occasionally a city slicker objected to following the "wanton waste" rule in Alaska's fish and game laws that forbids a hunting party from sawing off the trophy of a moose's antlers and leaving the meat to rot.

"So what you gotta do," Rick told me as we cruised over a broad plain beyond the riverside mountains, "is loosen the moose's tenderloin from the back and be careful not to ruin it by cutting it in two. Then you separate the ball joint from the legs and remove the ribs, the front shoulders and the rack. Now, the heaviest load is going to weigh about a hundred and twenty-five pounds and it will take a few trips to pack all of the moose out. So what you're going to have to do when you have the whole thing butchered is tie the first load on your pack, crawl under it, and then slowly get on all fours, and then hopefully you'll get upright."

In addition to chasing moose, Rick was part of a duo that was adept at finding productive "leads"—areas of open water amid the ice that are typically full of seals. These are places where polar bears like to congregate. This was before the passage of the 1972 Marine Mammals Protection Act, which made the "take" of polar bears and other mammals illegal, except for subsistence hunting by Alaska Natives. Back then, at the height of the Cold War, there were plenty of polar bears. On one occasion Rick got slightly turned around and started following a lead westward instead of eastward. When he looked up from the blinding whiteness of the pack ice he saw land he didn't recognize and realized he had flown into Soviet airspace. He hightailed it back safely.

His free enterprise career as a big-game-hunting guide, commercial pilot, and small-business operator helped his campaign to win a Republican seat in the Alaska state legislature. And his cautious yet adventurous style helped him solidify his political position, and serve in the house and senate for over twenty years, leading eventually to his becoming the senate's president. There he rubbed elbows with many of the state's Republican elite, including Sarah Palin. "She was one of the people that I consider a friend. Before she was chosen by the McCain campaign she did all the right things. She got the oil companies to double the amount of revenue they were paying to the state. She took on the corruption in the local Republican Party. But when she was selected by McCain, she became a mouthpiece for the party and all that vanished by the wayside."

But Halford did not stray from what he saw as the basic principles of the Republican Party, which happened to be mirrored in the twin poles of his character: caution and adventure. And indeed it is with an eye to both caution and adventure that he has lived his whole life. A dutiful father to three daughters by his first wife and very recently, at the age of sixty-six, the new and still very dutiful father to three sons by his second wife (a native Yup'ik Eskimo), he understands that his survival is the key to his family's survival. Which is how he came to take an interest in the plans for Pebble Mine, something that, in his opinion, dangerously skewed adventure over caution.

"It was back in 2006. It was winter and I had fairly recently retired from the legislature. I was headed for our home in Aleknagik and I was only about thirty miles from home," Rick told me over the headsets as we banked and headed northeast, the almost savannalike territory below speckled by shimmering lake potholes and brown grass where grizzlies, moose, and caribou surely loitered. "This is, of course, the most dangerous time to fly. You really want to get home and you'll fly through anything to get there, but this is exactly the time you should say to yourself, 'Wait a minute—I need to take a break.' My windshield was entirely iced up and I could only see through about a six-inch hole in the windshield from the defroster."

"Anyway," Rick continued after a bump dropped us a few dozen feet landward, "I saw a small village below me and decided I better land there. That's how I ended up in the village of Ekwok. I didn't know anyone there so I knocked on the nearest door, where Buck Williams, who is also a guide, gave me a place to stay for the night. The next morning, Buck invited a couple of people over for breakfast who were interested in talking to me, including Luki Akelkok, a Yup'ik fisherman, guide, and local community leader. They knew who I was because they'd seen my face on the rural public television network subsidized by the state. It's one of the only stations they get, and it airs state legislature proceedings.

"At breakfast Luki sat me down and told me that there was something going on with the prospect up near Frying Pan Lake. He had a cabin just upstream of the site and was nervous about the activity. I had been involved with resource issues for most of my career in the legislature. And I had a lot of experience with mining. My experience to date had been with small-scale mining companies that didn't do much damage and in fact had often done a lot of good. But every little prospector has a dream of hitting the mother lode. I don't think anyone realized the size of this thing."

What Luki Akelkok told Halford that chilly morning in the village of Ekwok echoed something he'd heard only in passing years earlier from Bella Hammond, the wife of Alaska's governor at the time. But when he heard more details from Akelkok he realized something was being conceived for the region that was of an entirely different scope than other projects he'd come across. For a long time, as far back as the early 1970s, mineral specialists had known that gold and copper existed beginning a hundred miles northwest of Bristol Bay and extending southeast down into British Columbia. The Bristol Bay strike's discovery was piecemeal at first. A hole here, a hole there. Oftentimes prospectors didn't pay it very much mind. Yes, there was copper and gold and even some molybdenum, an element used for the oddly divergent purposes of taking mammograms and fertilizing cauliflower. In nearly every hole dug by Cominco, the company that first laid claim to the strike, there was always some mineral of value. But it was extraordinarily diffuse. As developers state today, less than 3 percent of the rock in the area contains anything of value. Moreover, the land that the strike lay under was tricky. Not only was it wet and marshy, but there were considerable bureaucratic hurdles to overcome to lay claim to it. Governor

Jay Hammond, a legendarily quixotic figure who was as fond of writing doggerel verse as he was of weight lifting, had seen the long-range threats to the salmon and trout fisheries from mining and industrial development and as a state senator had attempted to protect them through passage of the 1971 State Congress House Joint Resolution No. 16 and the 1972 estabilishment of Bristol Bay Fisheries Reserve. In 1984 under Governor Bill Sheffield, the state adopted the more expansive Bristol Bay Area Plan. This plan loosely designated the state-owned portion of the bay's uplands as primarily salmon and wildlife habitat, a further disincentive to miners.

And indeed the overall cost of accessing the ore in the rock north of Dillingham was prohibitive. The nearest road to the lower forty-eight was two hundred miles away. True, there was an outlet to the sea thirty miles south on Iliamna Lake, a huge freshwater body nearly the size of America's larger continental Great Lakes. But there was barely a road along Iliamna, no fuel depot, no way to possibly move that much ore out to barges on the sea.

Still, by 1988, Cominco believed it had a strike of great value on its hands. Investment in exploration increased from 1988 to 1997 until, owing to "environmental concerns" and the expense of extracting the low-grade ore, Cominco didn't want to take the strike further. In 2001 Northern Dynasty, a Canadian company, bought in. A little while afterward they were joined in the endeavor by Anglo American, PLC. Headquartered in London, Anglo American is among the largest mining companies in the world, with assets in excess of $66 billion. It is understandable that such large interests would be attracted to the Pebble site. When the accountants ran through the potential of the deposit and the cost of exploitation, the math worked. In all the strike was valued at $500 billion.

As the value of the strike came to be understood, the politics surrounding it also changed. In 2005 Governor Frank Murkowski appointed a former mining executive as head of Alaska's Department of Natural Resources, and soon the Bristol Bay Area Plan was revisited. Land that the Area Plan had designated primarily as salmon habitat was re-designated primarily as mineral land, making it easier to mine. And finally, with Anglo American and its partners in place, a base camp was established just north of Frying Pan Lake in the highlands north of Lake Iliamna.

For Rick Halford, this would normally have been a good thing. He often boasts that during his tenure in the Alaska state legislature "I never came across a mining project I didn't like." But when Halford flew over the area and then researched the facts and dangers of the strike he realized that this was something very different. If developed, it would entail the extraction of over ten billion tons of ore.

"How big is ten billion tons?" I wondered aloud as our plane neared Frying Pan Lake.

"Well, I asked an engineer to put that in physical terms for me," Halford said. "It turns out that would be a three-dimensional rectangle a thousand feet wide, a thousand feet high, and twenty to thirty miles long." A thirty-mile-long block of sulfurous material in some of the most pristine country in the world, for it turns out that the Pebble deposit has much more sulfur than copper—as much as 7 to 8 percent is sulfur. As Halford puts it, if you were to describe the mine by what it really is rather than by what it tries to be, it would be a sulfur mine.

"If they go forward with the mine on the scale they are saying they want to," he continued, "down there at that mountain, that's where one end of the hundreds-feet-high dam would have to go in for the retaining lake. And all the way over there, several miles away, that's Sharp Mountain, that's where they'd probably have to attach the other end. The bad thing is that the dam and the lake would sit on the surface divide between Upper Talarik Creek and the Koktuli River, both fishing rivers. And both have underground connections to the two major watersheds in the area, the

Kvichak and Nushagak Rivers. You couldn't pick a better place to ruin both drainages."

I let my eyes wander over the sweep of tundra, the wildness, the scope of land and water that would be encompassed by this. I had read about projects like this—the mighty hydroelectric dams that were thrown up on the Columbia River in the 1940s, destroying the biggest runs of salmon in the continental United States; the miles upon miles of oil infrastructure laid beneath the Mississippi delta from the 1920s through the 1970. Now I was looking at something of that scope in its earliest stages. It was otherworldly.

And then, arising out of nowhere at the edge of the wilderness, there appeared below us a jagged little encampment. A helicopter swirled about; tiny dots of workers moved below. No actual mining was going on yet. It was only a station for exploratory holes. But the Pebble Partnership has already made many of them. "So far all the mining interests have drilled over twelve hundred holes, some up to five thousand feet deep," Halford said. "They have probably connected all the aquifers. There's a good chance they've already done a lot of damage."

The aquifer is the lifeblood of the hills and plains around Bristol Bay. The headwaters of the Kvichak and Nushagak

Rivers flow around all the copper and gold that the Pebble Partnership would lay claim to. And in those headwaters are the spawning grounds of some of the largest salmon runs left on earth. Enough lox for every bagel in New York City and the product, by the way, of tens of millions of years of evolution.

The Dawn and Dusk of Salmon

The oldest salmon-like creature we have found to date appeared about forty to fifty million years ago in the Eocene epoch. The Eocene, whose name comes from the Greek *eos*, "dawn," was quite literally the dawn of many of the preconditions that allowed human beings to exist. At the outset of the Eocene, carbon dioxide in the atmosphere spiked at levels well over 1,000 parts per million (our current CO_2 concentrations are about 400 parts per million). No large mammals walked the earth then and the poles were at times downright balmy. By the end of the epoch, global temperatures would begin to cool and approach temperate levels, allowing for the rise of the precursors to our modern cultivated crops. The oceans' carboniferous bivalves and plankton would surge and sequester carbon at a fantastic rate, bringing carbon in the atmosphere down to around 700 parts per million. Swarms of krill would begin to form around aggregations of plankton in the cooling waters of what would become the Bering Sea. The poles would take on their present form, cold and icy, and a regulating mechanism for the modern climate. It was right in the middle of this transition that salmon began to emerge.

The several different Pacific salmon species evolved life cycles that brought them to a stretch of ocean from the Bering Sea to the California coast, where plankton and krill converge. Sockeye salmon are particularly dependent on krill. And by feeding on krill, sockeye salmon acquire a plankton-synthesized compound called astaxanthin, a red-orange pigment that gives sockeye salmon flesh its characteristic ruby hue.

When we eat sockeye, we too acquire astaxanthin, an antioxidant and possible reducer of inflammation. We also get high levels of heart-healthy omega-3 fatty acids. A single three-ounce portion of cooked wild sockeye salmon contains eight hundred milligrams of omega-3s. To give an idea of how rich that is, physicians recommend that individuals suffering from heart conditions consume what they consider a large dose of omega-3s: eight hundred to one thousand milligrams per day.

This nutrient-loaded fish is the money salmon of Alaska,

and in Bristol Bay it is intrinsically dependent on the natural workings of the river and lake systems where it is born. Elsewhere in Alaska, particularly in the southeastern part of the state, large investments have been made in commercial hatcheries to maintain the abundance of salmon runs. Critics call this practice "salmon ranching" and fear that it may be skewing the genetics of those once completely wild runs. But no such hatcheries prop up the sockeye populations of Bristol Bay, making the Bristol Bay sockeye perhaps the world's most significant, purely wild commercial salmon run.

One reason hatcheries are a smaller component of the sockeye's life than that of other salmon is the species' difficult-to-master life cycle—a life cycle that intertwines it with the lake-rich inland areas of Alaska, British Columbia, Washington, Oregon, and northern California. Within the sockeye population subsets of the species spend varying amounts of time in ponds and lakes. Other salmon species generally move quickly out of their natal rivers, anxious to "smoltify" and transform into silvery, salt-tolerant sea goers. But some varieties of sockeye linger in lakes for much longer—sometimes as long as three years.

The sockeye's lake fixation has held it in good standing for many millions of years. It gave the sockeye an inland option when environmental conditions made quick migration to the open ocean disadvantageous. And by far the most important expression of the Bristol Bay sockeye's lake habit is Alaska's biggest lake, Iliamna, a lake that at any given time may contain more than a billion sockeye, either those lingering before smoltifying and heading out to sea or those catching their breath before they make their final push to spawn and die. Hundreds of millions more salmon tarry in nearby Lake Clark, another Alaskan "great" lake essentially unknown to most Americans.

If the Pebble Partnership has its way, the shores of Iliamna could become the site for a conveyor that would transport about three hundred million tons of minerals heading out to barges in the bay. Upstream some kind of retaining facility would have to be built—most likely a holding pond that would need to be able to contain ten billion tons of mining tailings. If an earthquake were to hit the region, pollution from those sulfurous tailings—along with the other chemicals that are typically used for gold extraction—could enter both watersheds, damaging first Iliamna Lake and then Lake Clark.

The possibilities of an earthquake are quite real. The Pebble strike sits on the Ring of Fire, the arc of seismically active volcanoes that circles the Pacific Ocean. Many fault lines run through the territory, and a total of seven earthquakes were recorded in 2012 alone. One fault line is just ten miles away from Frying Pan Lake.

That all this should be taking shape today in Bristol Bay of all places is heartbreaking. Because it is only today, after a century of conflict and bad fishing practices, that fisheries biologists and fishermen have finally struck what seems to be one of the best balances between fish and fishermen known in modern times. For the first hundred years of human-salmon interaction in Alaska, the same old story of overexploitation had been very much the rule. As Carl Safina writes in his excellent book *The View from Lazy Point*, Alaskan canneries in the 1800s "deployed nets a mile long and erected permanent barricades that took every fish attempting to enter a stream. For them there was no concept of 'enough.'" But then something unusually promising happened. After President Eisenhower declared the Alaska salmon runs a federal disaster, Alaska achieved statehood and banned large-scale salmon traps. Since then, sustainable fisheries management has actually been a tenet of the state constitution.

And now, with state-of-the-art fisheries management technology in place, with sonar counters, with satellite telemetry, with millions of dollars at their disposal to make sure that the hundreds of millions of dollars' worth of Alaska sockeye continue to return to those Alaskan rivers, suddenly the rug could be pulled out from under the whole complicated balancing act.

How did we reach this salmon precipice? Was it not enough to have destroyed the Atlantic salmon runs of New England when dam after dam was constructed throughout the tributaries of that once great salmon country? Was it not enough to have eviscerated the Pacific salmon runs of California, Oregon, and Washington during the New Deal era when dams were constructed on the Columbia, Snake, and many other Northwest rivers, and runs of many millions of fish were reduced by an order of magnitude? Were we not adequately forewarned when the *Exxon Valdez* ran aground in Prince William Sound in southeast Alaska, oiling the very estuary where so many salmon were hatched? Apparently not. Now the ultimate test is being offered to American rationality.

Bristol Bay is the most valuable salmon fishery in the world, generating $300 million in local profit and $1.5 billion when all the affiliated business it creates is taken into account. But the value of the fishery is dwarfed by the value of the mine, at least in the short term. The mine is valued at potentially $500 *billion*. True, amortized over a few hundred years, the salmon are more valuable, but you cannot build a shareholder base over a few hundred years.

The Pebble Partnership points out how environmentally

sound its development will be. The project planners maintain that Pebble will be smart mining. And it is generally hoped that the spills and failures that happened in the Bingham Canyon Mine in Utah, which contaminated sixty square miles of aquifer and trout country twenty years ago, will not be repeated. But the Pebble Partnership does not deny the sheer size of the mine. Billions of tons of tailings will come out of the mine and will require perpetual on-site remediation—in other words, environmental oversight forever. The Partnership does not deny that they are seeking to put an extremely large piece of industrial infrastructure within the context of a food system that has a very low threshold for pollution.

How did it come to pass, then, that what could be a tremendous source of brain-enriching nutrition is being subjected to so great a risk?

The answer lies in two lines of economic reasoning. The first is the way that fish wealth concentrates itself, or rather the way it fails to do so. A big minerals strike—be it oil or gold or molybdenum—represents a large, condensed chunk of wealth that private shareholders in a large, publicly traded company may acquire. Significant investment is required to convert that rock or petrochemical into money, but those sorts of investments are made by companies whose CEOs are accustomed to risk, knowing full well the size of the reward.

Fishing offers no such payoff. Yes, the Bristol Bay fishery may be worth well over $300 million a year, but unlike corn or wheat, wherein large, politically connected farming conglomerates now dominate the industry, Alaska salmon wealth is a diffuse and democratic wealth, spread among Yup'iks and whites and canneries and canny captains. Many people who work the fishery make a decent living. It is not unusual for a captain to bring down a hundred thousand dollars in a season. But what is a hundred thousand dollars when compared with a mining executive's salary? There is no way for the power of fishermen to concentrate itself enough to combat the wealth of mining and energy companies when conflicts arise.

But the other factor that distracts us from protecting our own seafood security is much more pernicious and counterintuitive.

Quite simply, Americans are risking their wild salmon because Americans do not eat enough of their wild salmon.

The people who love to eat wild salmon *really* love it. But most Americans are not wild salmon lovers. They find wild salmon too, well, salmony. Two-thirds of the salmon Americans consume is farmed and imported. And it is this preference that has masked the disappearance of wild salmon from our plates. If we like the taste of farmed salmon better than wild salmon, what need do we have of clean, unimpeded rivers and all the other environmental preconditions that make wild salmon possible? If we never see a paucity of salmon at our supermarkets, why should we concern ourselves with endeavors like Pebble Mine?

This all became clear to me during a seafood investment forum I attended not long after my return from Bristol Bay. The prospective investors had gathered in the ballroom of New York City's Essex House hotel, called forward by the seafood industry news agency IntraFish. The proceedings began bullishly, with Henry Demone, the CEO of one of the world's most profitable seafood companies, High Liner, noting that "now has never been a better time for seafood." It was put forward during his and other presentations that the world by 2030 would need forty million more metric tons of seafood if the rise in demand was going to be met.

But then the nature of where all this demand was coming from was revealed. Joe Bundrant, the chief executive of Trident Seafoods, a major trader in Bristol Bay sockeye salmon, took the stage and, like the rest of the bullish investors and executives in the room, continued the thought—on how much fish was going to be sold, how good the future looked. But where was all this growth coming from?

"What's the biggest opportunity out there?" Bundrant asked rhetorically. "In a word, China." It turns out that it is the rising middle class in China that is most interested in seafood. It is the Chinese who insist upon whole fish, fresh fish, fish that is, actually, quite fishy. And while Americans wring their hands about sustainability, about what they should eat or whether they should eat fish at all, the Chinese are surging ahead. Today 79 percent of all Alaska salmon is exported, and an increasingly large share of that is going to China.

"There's a saying," Colin MacDonald, cofounder and chairman of Clearwater Seafoods, said as he took his turn at the dais, "Americans eat with their minds. The Japanese eat with their eyes. The Chinese eat with their mouths." And China has quite a few mouths—so many, in fact, that they endanger our own food security. In September 2013, Shuanghui International, one of China's largest meat processors, bought Smithfield Farms, the United States' largest pork producer. And while Chinese imports of American fish are still minor in comparison to Japan's, they are rising and the major U.S. seafood exporters are taking note.

Chinese acquisition of American food infrastructure points to another factor about China that is driving all this demand. Something that is embodied in another saying popular in China today: "If you eat the food in China, that will kill you. But if you do not eat the food in China, that will kill you even faster."

After it was revealed in 2007 and 2008 that Chinese pet

food, milk, and aquaculture products were tainted with the carcinogenic additive melamine, a major outcry took place both in China and among importers of Chinese products. Zheng Xiaoyu, the head of China's State Food and Drug Administration, was executed in 2007 for accepting bribes that are acknowledged to have been linked to much of the early failure in inspection. After Zheng's execution, the inspection process for Chinese exports has become more stringent. It is today much less likely that Chinese export products—all that tilapia, scallops, shrimp, catfish, and redfish sent to America every year—present a real danger to American consumers. Nevertheless, fears remain, and country-of-origin labeling doesn't necessarily help. Fish fillets must carry a label indicating where that fish came from, but highly processed imported seafood—fish sticks, crab cakes, for example—do not.

And what of Chinese consumers? What assurances do they have that the food that stays locally in China is safe? Many in China remain wary of their own food. But whereas previously not eating the food in China could indeed kill you—there was nothing else to eat—the rapid rise of a middle class and the opening of Chinese markets to foreign goods means that suddenly there is a way for Chinese who do not want to eat the food in China to eat something else. This is supported by the Communist Party itself, which since the 1950s has maintained the so-called *tegong* system for its elites. Meaning "special supply," *tegong* provides a completely separate, healthier, and sometimes even organic farm network for upper-level Communist Party officials. Currently there is no special provision within the *tegong* for American fish, but it is easy to imagine it happening. With its wildness enshrined in the global cultural zeitgeist (such is the influence of the much translated and widely read Jack London), Alaska can trade on its reputation for being clean and wholesome. The Chinese, like everyone else, would like to eat food that is safe and nutritious. The nearest place with large quantities of clean and nutritious seafood is Alaska.

This reveals the less widely known side of the much publicized American seafood deficit. It's often mourned in American governmental circles that 91 percent of the seafood Americans consume comes from abroad. But what American seafood deficit hawks are not as quick to mention is that the United States exports many millions of tons of seafood. Two-thirds of all Alaska seafood, a good chunk of which is salmon, is sent abroad. And much of the Alaska salmon that does make it to American consumers is flash-frozen whole, shipped to China, defrosted, filleted, and deboned. It is then refrozen and shipped back to the United States as twice-frozen boneless product.

Nowadays, however, an increasingly large part of the

seafood we send to China stays there for Chinese consumption. This trend is augmented by the dissolution of the American fish-processing industry. From scallops and haddock in New Bedford to spiny lobsters and black cod in central California, I found again and again in my research that fishermen had poor access to selling and processing their fish locally. In American coastal communities the fish business has been pushed off the dock by the real estate business and as a result it is much easier and cheaper for fishermen to send their catch, unprocessed and in bulk, to China. There it finds cheap labor to turn it into edible food and a hungry fish-loving nation ready to consume it.

How bad a deal all this moving around of fish is for Americans became clear to me at the IntraFish conference in New York. After the flood of good news about Asia quieted down, the conversation turned to the gloomy American side of things and the way Americans turned up their noses at food from the sea.

"When we do focus groups, there are three concerns that come up from Americans," a panelist said at the outset of another investors' roundtable. "When Americans are asked why they don't want to eat fish, they say: I don't want to touch it, I don't know how to prepare it, and I don't want it smelling up my kitchen."

In this, there seemed to me a tremendous opportunity.

The seafood industry could see American ignorance about fish as an avenue for education—a mass campaign focusing on fish's health properties, with lessons on how to properly purchase, prepare, and cook seafood. But how did the seafood industry respond at the IntraFish conference?

"We've developed semipermeable microwave film to create microwave products!" one of the panelists boasted, and then, to back up the claim, a series of advertisements was shown of fish products specially targeted to the American consumer. Preseasoned fillets sealed in "semipermeable microwave film" that can be popped into the microwave. The film miraculously puffs as it's cooked and then needs only to be cut open once the fish is on a plate on the table. No smell in the kitchen! And what kind of fish were in these precious little packets of semipermeable microwave film? The least fishy fish of all: tilapia. Farmed largely in Latin America and China, tilapia is odorless and virtually tasteless, with very little omega-3 content.

It was there and then that it hit me—the bizarre devil's bargain that Americans have entered into with their seafood supply. Americans now harvest our best, most nutritious fish in our best-managed Alaskan fisheries and send those fish over to Asia. In exchange, we are importing fish farmed in Asia, with little of the brain-building compounds fish eaters are seeking when they eat fish. This fish can be difficult to trace back to its farm of origin, and less than 2 percent of it is inspected directly by the FDA. In the case of shrimp, that food is higher in cholesterol than most seafood; in the case of tilapia, instead of heart-healthy omega-3s, it is rich in omega-6s, which are not as beneficial and may in fact be harmful when consumed in large amounts. Many physicians argue that high omega-6 fatty acid content—more typical of landfood like beef and pork—is detrimental to human health and can aggravate heart conditions when poorly balanced with omega-3s. But because American consumers don't really know their own fish, they tend to lump it all together under that word "seafood" and presume that it is all basically "good."

And so a frustrating disconnect has emerged on both the foreign and domestic sides of American fisheries resources. Foreign buyers would like the salmon they buy from Alaska to continue to be wild, clean, and readily available. But foreigners have no jurisdiction over projects like Pebble Mine and the industrial development that could eradicate the food that they most want. Americans, meanwhile, who eat seafood raised on Chinese farms would like that Chinese seafood to be grown without the use of carcinogenic chemicals and raised in clean, safe water. But we have no control over the Chinese food system. We can demand standards of our importers and certification of the supply chain, but the rules and regulations that govern Asian aquaculture are local and hard to assess. Until Vice President Biden brokered a recent agreement, the Chinese government had been actively denying visas to additional FDA inspectors the United States was trying to place in China. And while even America's largest seafood advocacy organization, the National Fisheries Institute, maintains that China has made great strides in food safety and that Chinese seafood is as well inspected as seafood of American origin, there is an opacity to the Chinese system that leaves ample room for speculation. Air quality on an average day in Beijing has been regularly recorded by the U.S. embassy as being worse than Manhattan's a few days after 9/11. If our direct monitoring is picking up this kind of egregious pollution in the most observed part of China, can we really expect that water quality and aquaculture feed in regions far outside the observation posts of foreign embassies would be markedly better? According to a recent *New York Times* op-ed by the scholars Damien Ma and William Adams, more than half of China's largest lakes and reservoirs were so contaminated in 2011 that their waters were unsuitable for human consumption. This according to the Chinese goverment's very own standards.

All this makes you want to stamp your feet and scream out loud, to shout at Alaska to just stop. Stop sending your fish abroad. Stop your plans to ruin your rivers. Focus on what you do best. Feed America your good food.

Instead Alaska does not respond to these kinds of entreaties. A campaign has to be waged to remind Alaska that it is not a country in its own right, that it is a state in a nation that has environmental rules and regulations that were created for the betterment of us all.

Fighting Back

"Now what's this, young man?" a slightly indignant older woman said as she reached up, grabbed the brim of my fishing hat, and yanked it down over my eyes so that I could barely see my surroundings—the West Conference Room of the United States Supreme Court.

I was due to give a brief talk about Pebble Mine that evening in the court at a reception set up by the nonprofit Wild Salmon Center. Over the last year of lecturing I'd grown accustomed to wearing my fishing hat whenever I spoke publicly. Part of it was habit. Part of it was branding—I was the fish writer who lectured wearing a fishing hat. I'd actually considered not wearing my fishing hat at the U.S. Supreme Court—it was the Supreme Court, after all. But then I'd heard the story of how Chief Justice William Rehnquist had begun wearing gold stripes on his robe because he'd seen a production of Gilbert and Sullivan's *Pirates of Penzance* in which a judge had worn gold stripes. Inspired by the fictitious judge, the real judge thought that gold stripes would give him a certain élan, so he had them sewn onto his official robe. I figured if Rehnquist could wear fake gold stripes in the Supreme Court, I could wear my real fishing hat. But the older woman was scowling at me so authoritatively that it seemed like a good idea to reconsider.

"But this is my fishing hat," I said finally.

"Hmph," she said, and continued to scowl.

"Would you like me to take it off?" I asked.

"Yes," she said. "Yes, I would."

I did so and she immediately brightened into the practiced, radiant smile of someone who had risen high in public office. I blushed hard and tried to recover my poise.

"I'm Sandra," Justice Sandra Day O'Connor said, jutting out her hand. "And I'm a fisherman too."

Beside me the pilot and Alaska state senator Rick Halford patted me on the back. "When the judge says to take off your hat," he laughed, "you better take off your hat."

There is an old Russian expression that often comes to my mind at moments like this: "A fisherman can spot another fisherman from far away." Regardless of caste, religion, or political affiliation, people who really like to fish and who really like fishy country are united in a loose confederation. They are buried like Masons or CIA moles throughout government and industry. They mostly keep the existence of their secret order to themselves. But push a fisherman too far, go into his country and screw around with his river, and others will appear from out of the woodwork.

Such was the case on this April evening in Washington, D.C. Sandra Day O'Connor had fished Alaska on many occasions and was clearly of the same Republican stripe as Rick Halford. The old-school, Teddy Roosevelt–type of Republican. O'Connor had arranged for the Wild Salmon Center reception at the court. Through her minders we were asked not to speak of Pebble Mine that evening—only to speak about the glories of Bristol Bay and the salmon that swam there. But an odd assortment of fishy characters had come into the court building to give their support and were ready to vocalize should they be allowed to do so. A representative of the jeweler Tiffany and Co. was there (Tiffany's CEO lives on the Jersey Shore and has a deep love of the coast). Tiffany had pledged not to source gold from Pebble Mine should it be built. Stone Gossard, the guitarist of the rock band Pearl Jam, had come and now stood in line for the microphone with me. And Lisa Jackson, head of the EPA, was there too. Though markedly noncommittal and clearly under instructions from the Obama administration to remain as neutral as possible on Pebble Mine, she had previously served as the New Jersey commissioner of environmental protection and seemed to know the value of fish.

This unlikely posse had gathered because of an equally unlikely discovery that had been made the year before and brought to the attention of Shoren Brown, an environmentalist working for the nonprofit Trout Unlimited on the Pebble Campaign. As I wrote earlier, the Pebble deposit sits on land that belongs to the state of Alaska. The federal government has no jurisdiction over it at all. But when several different environmental groups in Alaska began trying to find some kind of reasoning—anything—that could possibly let the rest of the country weigh in on the future of America's most valuable salmon fishery, they stumbled upon something that was a game changer.

Section 404(c) of the Clean Water Act authorizes the EPA "to prohibit, restrict, or deny the discharge of dredged or fill material at defined sites in waters of the United States (including wetlands) whenever it determines . . . that use of such sites for disposal would have an *unacceptable adverse impact on one or more of various resources, including fisheries* . . ." Somewhere deep in the past of the idealistic 1960s and 1970s when the Clean Water Act was drafted, somebody had predicted something as big as Pebble would come along.

It is particularly fitting that it would be Shoren Brown who would lead the national campaign that has relentlessly pressured the Obama administration for years to use this seldom-used tool to stop Pebble Mine. Shoren grew up in southern Louisiana in the same Cancer Alley that has polluted so much good shrimp country. Indeed, if you ever were to go looking for a place where section 404(c) of the Clean Water Act *should* have been implemented, the backwoods of Louisiana, which the petrochemicals industry had so degraded for nearly a century, should have been that place.

But in spite of the intensity of the pollution, the region was also rich in fish and game, which were the loves of Shoren Brown in his youth. "There was nothing—and I mean *nothing*—I liked better growing up than disappearing into the bayous of Louisiana and fishing," he told me. He had somewhat quixotically grown up on a hippie commune (hence his earthy, crunchy name), and getting into the woods had been the only way for him to escape the cloying hugs of his tie-died communard parents and their hippie coparents. Fishing in the woods, Shoren was able to blend in with the good ol' boy Cajun neighbors, and was given a proper Southern nickname: Shorty.

I immediately liked Shorty Brown. I too used to disappear into the woods for days at a time on fishing trips and in my youth would have rather been fishing than doing pretty much any other thing. In the months Shoren and I came to know each other a quiet whisper of competition had developed between us as can only develop between fishermen. When we traveled together to Bristol Bay, that competition bloomed into all-out warfare. The rules of a fishing contest in most humdrum, depleted fisheries are usually pretty straightforward. Biggest fish usually decides a winner, and if that doesn't work sometimes competitors will resort to most fish. But in Bristol Bay you have to go a good deal deeper to come up with meaningful contest. A biggest-fish competition doesn't work—salmon are generally all the same age when they come into the river in their many millions and thus are more or less the same size. A most-fish competition doesn't really work either. If you were to hit a big run of sockeye, you could fish until your biceps popped and you'd lose count just about the time you lost consciousness.

No, the only reasonable competition in this setting is the ultimate quest for fishermen around the world: total number of *species*. And herein Bristol Bay was equally astounding. For the week that Shoren and I fished around Bristol Bay, we found ourselves deadlocked. We'd each caught lake trout with their psychedelic squiggle-patterned flanks, arctic grayling sporting their sail-like fins, chum salmon, king salmon as big as German shepherds, and northern pike with teeth that could slice a human finger in half. For a little while we lingered on rainbow trout—that fish being the great prize of the sport fishermen who journey to Bristol Bay. Salmon are too numerous to get wealthy anglers excited and to flatter their entitled sense of exclusivity, so the superrich who fly in to Bristol Bay on private jets and stay at thousand-dollar-a-night lodges come for the giant rainbow trout that feast upon the rotting salmon carcasses. In most other places when you fly-fish for trout you try to "match the hatch"—i.e., choose a fly that matches the particular midge or mayfly that might be emerging from the water's surface. But in Bristol Bay the giant trout don't bother with insects. One of the most effective "fly" patterns for a Bristol Bay rainbow trout is what's called a "flesh fly," a droopy, unattractive mass of down feathers that is styled to look like a chunk of rotting salmon carcass. That, or an "egg pattern," meant to mimic the oily globs of salmon egg that the big rainbows also wolf down with abandon. That's what makes the Bristol Bay rainbows so big and mean. They cram their bodies with the nutrients the salmon have brought back from the sea. And for a day or two in Bristol Bay the rainbow trout competition between Shoren and me was intense. Shoren caught a twenty-three-inch rainbow. I caught a twenty-four-inch fish—half again as big as the largest rainbow I'd ever caught in the East. Shoren then caught a twenty-six-incher. And then fishing off the pontoon of the floatplane one evening I'd caught one even bigger. When our guide came over to hold the fish up for a picture, the rainbow leaped into the air out of his hands. The guide, trying not to hurt the fish, caught it, but then it slipped away again and with its tremendous salmon-flesh-fueled power plant of a tail shot five feet into the air. Again the guide caught it, only to have it slip away a third time. All the raw energy of the place inhabited this fish and the crimson stripe down its center glowed in anger. Again it was caught for a picture and again it slipped away. Finally, after a few more attempts it was eventually cradled and calmed. Shoren laughed at the fish as the guide slipped it into the water. In the tamed, beaten-down waters of my home state of New York such a fish would have waddled pathetically in the current and probably would have died. But here in this most healthy of places the fish was still raw quicksilver. It took two short breaths and then shot away against a current that was ripping in the other direction. With that the rainbow trout competition between me and Shoren effectively ended.

Like that angry rainbow trout, Shoren has been dogged in his fight on Pebble Mine partly because it reminds him of so many lost fights in Louisiana. "Alaska and Louisiana have a lot in common," he told me. "They are both resource states—places where big extractive industries have a lot of political sway and a long history of damaging fish and wildlife in pursuit of profit." In the five years leading up to his joining the Pebble fight, he'd helped in stopping the petroleum development of the Arctic National Wildlife Refuge (ANWR), leading grassroots efforts to help stall out that high-profile Republican-led initiative in 2005. But that conflict had been fundamentally different from Pebble. Nearly 75 percent of Alaskans actually support drilling in ANWR, and the proposition to open the area to drilling has always been supported by the entirety of Alaska's elected officials regardless of party. The strategy on that campaign was simple: engage elected officials in the lower forty-eight and get them to see that protecting ANWR was bigger than just an Alaskan issue. "And anyway," Brown continued, "in Congress it's much easier to play defense than offense. Which is why we won with ANWR. Sort of won, anyway." To this day the required legislation to begin drilling has stalled in Congress. But on the other hand, ANWR has not received legislative protection that would forever safeguard it from drilling.

With Bristol Bay and Pebble Mine, however, a rift had appeared within the usual alliance of the develop-at-any-cost bloc of resource users in Alaska and elected officials.

On one side of the rift were the old-style Republicans, like Rick Halford, who saw in the West in general and Alaska in particular the value of wildness to the American character as well as renewable industries that create jobs without devastating the environment. On the other side were the new-style Republicans, the politicians who felt that any government regulation smacked of East Coast liberal intervention. And it was into this rift that Shoren Brown shoved Section 404(c) of the Clean Water Act. Since 404(c) prohibits "the discharge of dredged or fill material" in areas where an "unacceptable adverse impact on one or more of various resources, *including fisheries*" is likely to occur, Pebble posed a huge problem. With no roads, no access to the sea, no contact at all with the outside world, what else can you do with a thirty-mile-long, one-thousand-foot high rectangle of sulfurous waste? The only thing you can do is leave it right where you found it in all its nature-threatening glory. This is precisely what section 404(c) was designed to prevent.

And so when I came to Washington and had my encounter with Sandra Day O'Connor I came to help make the Clean Water Act's 404(c) argument clear. I'd come to join Shoren, the pilot-turned-senator Rick Halford, Pete Andrew, a Yup'ik board member of the Bristol Bay Native Corporation (BBNC), and Bob Waldrop, the head of the Bristol Bay Regional Seafood Development Association. And argue we did. We argued in the West Conference Room of the Supreme Court that evening, while Sandra Day O'Connor did her best to both support the idea of a 404(c) intervention while keeping a judicial Olympian Republican air about the whole thing. "Now, I understand there's some kind of problem up there in Bristol Bay," was all she said in allusion to Pebble. But it was enough for all in the room that she had noticed the problem and had mentioned it publicly.

And beyond the court we also made the 404(c) argument before Senator Mark Begich of Alaska, a Democrat who has been receptive to hearing concerns about Pebble but who is reluctant to take action on the issue probably because of his tenuous Democratic status in a blood red Republican state. And lastly, we put the argument for protecting Bristol Bay to Sam Kass, a professional chef and now President Obama's food initiatives coordinator and executive director of the first lady's "Let's Move" campaign. As we sat down in the large oak-paneled room at the White House, Pete Andrew passed Kass a modest-sized jar of salmon he'd smoked himself and advised him to mix the oil in the jar with cream cheese after the fish was all gone. "Makes a nice spread," Pete said. Kass stared at the jar. "The other thing that's good," Pete continued, "is when the sockeye start to fall apart and rot toward the end of the run, their bones get soft and you can make a nice kind of half-crunchy thing. It's good on toast." Kass, clearly overworked by his job and probably longing for the life of a regular old chef, seemed transfixed by the idea of crunchy salmon bones on toast.

"I'd like to try that," I remember him saying wistfully.

But then, coming out of his reverie, he went back to the issue at hand. He emphasized that in the wake of the 2008 financial crisis everyone in Washington was worried about employment. What we really needed to do, I recall Kass saying, was focus on jobs.

"Oh, we've got jobs," Pete Andrew said with a smile. "We've got plenty of jobs."

There is nothing that reflects the strange combination of modernity and ancientness surrounding the Pebble Mine vs. sockeye salmon fight than sitting around in the middle of the night over Cokes and Sprites in a Connecticut Sbarro pizza restaurant and discussing international finance with a bunch of native Alaskan tribal elders.

In the next stop after Washington in the Pebble Mine campaign that Shoren Brown and the Wild Salmon Center had put together, I had gone to Yale University along with the native leadership of Bristol Bay for a panel discussion on

the mine and what it meant for the future of fish. We had
been joined at Yale via videoconference from Anchorage by
John Shively, the president of the Pebble Partnership. During the panel Shively had argued passionately on the mine's
behalf. And now, with the panel discussion complete, but
having missed our train back to New York, the Alaskan natives and I were sitting around in the Sbarro pizza joint de-
briefing, wondering aloud how it had all come to this with
John Shively and Bristol Bay. Some of the usual Pebble
Mine fighters were there—Shoren Brown was there, as was
Pete Andrew, once again with his jars of home-smoked
salmon. But this time a good chunk of the management of
the Bristol Bay Native Corporation had come as well. Chief
Joe Chythlook was there as chairman of the board, as were
vice chair Dorothy Larson and another board member, Russell Nelson. It was a sweet group of elderly ladies and gen-
tlemen with twinkly eyes and flannel shirts and baseball
caps. Some of them were reading bent-up old mystery pulp
paperbacks. Some of them would have looked quite comfortable with needlework in their laps.

And collectively they controlled assets worth $500
million.

The Bristol Bay Native Corporation is to my eye one of
the most unusual and perhaps one of the more interesting
reimaginings of American corporate life I've come across. In

1971, the federal government signed a truce with the indigenous peoples of Alaska in a landmark treaty called the Alaska Native Claims Settlement Act. In the settlement, different tribes were given large swaths of land in addition to cash, designed to help those tribes steward their lands. Since 1973 the results have been mixed from tribe to tribe. In some places quick deals were struck with oil and gas companies. But the Bristol Bay Native Corporation was savvy and salmony at the same time. Taking their initial $30 million grant, they immediately began diversifying. At first they bought the Peter Pan salmon cannery in Dillingham. They ran the cannery at a profit and then sold it to the Japanese. They then bought into the Sunny Jim peanut butter line of products, which they also flipped for a substantial profit. They then expanded into government contracting, taking on construction projects in the lower forty-eight as well as in Afghanistan and Iraq. Over the years they grew their original nugget of money tenfold to half a billion dollars. Throughout, they had never touched the three million acres of land they own in Bristol Bay. It doesn't even appear on their balance sheet. That land was sacred. And in terms of salmon, that land had been and continued to be profitable. It allowed a great variety of independent fishermen to work the estuary while still staying within the bounds of a sustainable harvest.

Like me, everyone on the BBNC board was deeply concerned about Pebble Mine and what it meant for their traditional way of life. A way of life that, while supplemented by
modern conveniences, still revolves around the salmon's annual arrival in the bay. A life that BBNC CEO Jason Metrokin said would be "village based and sustainable with a
paramount goal of protecting the salmon resource." But the
sweet old grannies and gramps of the BBNC board had
many things on their minds that evening in the New Haven
Sbarro. The following day they would be flying to Honolulu
to tend to some hotel properties they owned, but they had
also just gone to see their fund managers while in New York.
The managers must have been on their toes for the meeting.
"Last year one of them wasn't doing very well," Dorothy
Larson said sweetly. "He was not getting us a good return
on our investment. So we chose to end the relationship."

But as we sat chatting that night in the pizza joint, the
thing that disturbed all of them the most about the Pebble
deal was that they all *knew* the head of the operation, John
Shively. They knew him personally. Jason Metrokin was his
next-door neighbor in Anchorage. As they sat listening to
him at Yale, they were convinced that this man they had
known for years was not being straightforward in present-
ing what he and the Pebble Partnership had in mind. During the panel discussion at Yale Shively had tried to tell

them and the audience that Pebble was all just an idea at the moment. That they were just trying out the idea. That Pebble should be allowed to *present* their idea and be judged fairly for it. "I'm not probably what you think about when you think about a CEO of a mining company," Shively had said during the panel discussion, pencil and paper in hand, scribbling notes to himself whenever someone came up with a counter to his assertions. "I don't come out of the mining industry. I have been working with Alaska natives for seventeen years, including with the Native Corporation just like the one that Jason works for." He even admitted that choosing mining over fishing might be a bad idea. "If we are talking about a choice, then we don't mine. Because you would never in my mind make a choice of mining over fish. Fish are renewable, and they have been there a long time." But Shively maintained that nothing had started yet; this was no fait accompli. It was all just "research" at this point. The base camp I'd seen at Frying Pan Lake, the twelve hundred holes in the Bristol Bay aquifer, the Pebble Web site that boasted "The Pebble Partnership moves into its summer work season with a proposed $107 million program budget approved by its board of directors for 2012"—all just research. But Jason Metrokin was quick on his feet in the debate and had countered Shively adeptly.

"So, yes, this *is* about choice, John," Metrokin said to Shively. "But it is really about what the local people want, what the local people need. What confuses the issue and makes this challenging, very technical, and highly sensitive is that this proposed project is on state land. The state of Alaska owns the land where this project is being proposed. However, Bristol Bay, as it's made up, [has] multiple stakeholders and multiple landowners, including the federal government. There are a number of parks, refuges—both state and federal—and private landowners, including the Bristol Bay Native Corporation, which is the largest private landowner in Bristol Bay. We have three million acres that we own and protect . . . This is not the right project, it is not the right place, it is too big, the impacts would be so severe to the existing fishery."

When I had my turn to speak, I asked Shively directly if it was true that ten billion tons of toxic tailings would have to be stored on site and that it would require perpetual remediation. He somewhat deflected the question.

"The technology of tailings is ever changing," Shively demurred. "I think that, to be quite frank, that people don't know how much monitoring is going to have to take place with tailings. Certainly it's going to have to take a significant amount of time. And we will be asked to put a cash

amount of money up so that there can be perpetual monitoring as needed. I don't know how much money that is going
to be." How much money indeed? How much does perpetuity cost? How do you even go about reminding people ten
thousand years in the future that they are contractually ob-
ligated to pay for monitoring a mine that their great-great-
great-times-one-hundred-grandfathers built? And then I
also thought about another stakeholder that Metrokin didn't
mention: the American who could and should eat wild
American salmon. How were his interests being represented
in all of this? What guarantee was he being given that his
food would be safe and that the food of his great-great-
grandchildren (times one hundred) was being protected?

But Shively shied away from the issue of perpetuity, instead invoking that hallowed American tonic for happiness
known as jobs. "When I took this position," said Shively "I
took it because of the economic opportunities that I saw for
a region that is, to be point-blank, poor." After talking about
declining schools, exorbitant gasoline prices, and the value
of wilderness refuges, he continued. "I think that the wilderness part of Alaska is great, but it can't all be wilderness.
There has to be some opportunity for development and the
state [was] given the opportunity to take about a hundred
million acres of land that primarily was for development, so
that the people of Alaska would have economic opportuni-

ties. There's always tension in Alaska no matter what we do, whether it's oil fields in the north, potential oil fields off-shore, or the Pebble Mine—you know, how much wilderness can you have, and once you've determined that, can you have any areas that are right now undeveloped that can promote development and can promote opportunities for the people that live here?"

Perhaps Shively is right. Perhaps there are some places in Alaska that ought to be "developed." And Pebble does claim it will provide something like three thousand full-time jobs that will last until the mine is exhausted some eighty years from now. But why must the one place where wilderness is actually producing a sustainable profit for humans be the target of that development? It's true that what constitutes a seafood job in Bristol Bay is a fluid thing. Twelve thousand annual jobs are provided by the Bristol Bay sockeye salmon industry, but the work comes in bursts and the profits are unpredictable and fluid. Just how fluid is revealed the moment your floatplane touches down in one of the native villages and eases up to the clutter of boat parts, nets, random plywood, houses abandoned, houses half abandoned, moose meat, racks of drying salmon, a Russian Orthodox church or two, rolls of old wallpaper—everything you could imagine, not stored neatly in someone's cellar, but out there for everyone to see. As Bill Carter recounts in *Red Summer*, his excellent account of salmon fishing in Bristol Bay, there are no hedgerows or fences in these villages; a neighbor's property begins where the stuff of the adjoining property ends. And yet, in a decent native village where alcohol has not done its mischief and where a unity of purpose can be detected, the appearance of the salmon wipes all of the discord away, and the pureness of the pursuit and the quality of the fish suggest that there is in fact a lot of sense and order to the place. When the fish come in, they come in big and there simply isn't time to think about aesthetics.

It's important to remember too that when we talk about Pebble Mine we are not just talking about the risk to a handful of villages directly in the mine's footprint. Earlier in the year while I was in the native village of New Stuyahok fishing with the town's mayor, Randal Hastings, this came to me quite by accident. Hastings was telling me about how when he'd won the election for mayor he'd banned the sale of alcohol and gotten the town out of a ninety-thousand-dollar debt to the IRS. "But all of that could go down the tubes because of this Humble Mine thing."

"You mean Pebble," I said.

"No, Pebble Mine is on the west side. Humble is on the east. They're waiting for their permit too."

There are in fact fifteen other mining claims currently being explored in Bristol Bay. They are all queued up behind Pebble waiting to see what the EPA will do. If it does nothing and allows Alaska to continue with business as usual, Bristol Bay could end up much more of a minerals extraction system than a food system.

This weighs heavily on the minds of the Alaskan natives. It weighs on them in the boardrooms of Washington, in the oil fields of Iraq, at the Anglo American shareholders meeting in London. All of these places are haunts for the modern Eskimo now.

In the New Haven Sbarro, as the hour of our train back to New York approached, we tired of turning it all over in our heads. Eventually there was just silence, and a kind of sadness.

"The Pebble Partnership sure chose the right person for the job," Dorothy Larson said. "John Shively's a well-known figure in the state with ties to the state government as well as to another regional corporation and the Alaska Federation of Natives [AFN, a statewide native organization], as well as a former VISTA volunteer. I was surprised that John took the position. I have known him for years. We worked together at AFN. We just have to agree to disagree."

A Modified Future

I experienced one last political confrontation over salmon before I was able to leave the East Coast and get back to Bristol Bay and the salmon themselves. And even though it doesn't concern Bristol Bay sockeye directly, it is worth recounting here, for it is the thing that kicks the stool out from under the great salmon nation of the Pacific Northwest. As with shrimp in Louisiana, the issue that affects wild salmon the most is foreign aquaculture.

Farmed salmon appeared in earnest in 1970s and began surging, like shrimp, in the 1980s. Within a decade salmon farming had migrated from its original home in Norway and Scotland to Canada and Chile and elsewhere in the Southern Hemisphere. Soon, just as with wild Gulf shrimp, wild salmon prices fell precipitously. The fall in Alaska salmon prices was compounded in 1989 by the *Exxon Valdez* oil spill. Immediately afterward, Alaska fishermen all over the state—especially in the footprint of the spill—suffered at least a 30 percent loss in income. Many lost a lot more. American consumers, unaware of the use of antibiotics and colorants in farmed salmon, were more afraid of petrochemical pollutants even in areas where they hadn't been present.

Since the nadir of the late eighties and early nineties,

Alaska seafood has fought back from its trough. Aggressive marketing and branding helped. The opening of new markets in Asia helped even more. But now a new threat was arising from farmed salmon.

I had been contacted about this threat by the office of Alaska senator Mark Begich to see whether I might come down to Washington, D.C., to testify before Congress. Senator Begich and fellow Alaska senator Lisa Murkowski had decided to draft a bill that would ban the interstate commerce of a particular farmed fish for human consumption. The legislation around this fish had been, like Pebble Mine, lurking beneath the surface for quite some time—almost two decades. The fish had been discovered—or rather created—right around the same time that mineralogists at Cominco were coming to realize the enormity of the Pebble strike. This fish also represented a significant windfall set to be drawn from the privatization of a publicly held resource. But whereas John Shively and the Pebble Partnership were seeking to privatize something gigantic, this other entity, an entity called AquaBounty, was looking to privatize something microscopic. If the aquaculture biotech company AquaBounty got its way, it would release for public consumption the first genetically modified animal: the AquAdvantage salmon. The gene within it that made it grow twice as fast as any salmon on earth would be theirs and theirs alone. To date the FDA had not approved the AquAdvantage salmon, but it was about to. And this concerned me.

In a sterile side room of the Russell Senate Office Building, two senators who count salmon among their constituents—Mark Begich of Alaska and the famously independent-minded Olympia Snowe of Maine—sat high above us. To my right sat Dr. Ronald Stotish, CEO of AquaBounty, and mixed among us were various other academics and PhDs of various stripes.

Standing to speak, I shared my objections, namely that while the AquAdvantage salmon—an engineered fish fashioned from an Atlantic salmon strand of DNA joined with a Pacific king salmon growth gene and coupled with a regulator protein from an ocean pout—might be technically safe for human consumption, it was, above all else, an unnecessary risk. I noted the hundreds of millions of pounds of salmon produced in Alaska annually, and the fact that much of it went abroad. And even if we didn't have all that wild Alaska salmon, modern unmodified-salmon farms were getting increasingly more efficient.

But what I didn't say during the course of my testimony was that I had over the course of the last few years grown increasingly uncomfortable with certain aspects of aquaculture—in particular, the way fish and shrimp farming eroded our relationship with naturally reproducing American marine systems. When AquaBounty's Ronald Stotish repeated the usual arguments of the need for fish farming, I of course agreed with the larger metalogic. Yes, the world's population is growing. Yes, fish farming is a relatively efficient way to create protein—much more efficient, in fact, than the growing of land mammals. But what Stotish was doing in his testimony was something aquaculture entrepreneurs everywhere do: conflate global problems with American problems—problems that we Americans don't have . . . yet. We have no *real* shortage of wild fish in America. If Alaska continued as America's premier fishing place rather than becoming a resource extraction place, that problem would never evidence itself. The world might need a genetically modified farmed fish. But America does not. We still have a lot of wild seafood, and—if we were not trading it away—we could greatly reduce our imports. America, with a relatively low human-to-land ratio, could continue this way provided we kept our population in check and our coastal fish systems intact.

All this was going through my mind as AquaBounty's Stotish carried on with his testimony. "Global animal biotechnology is on the rise," he warned. "For instance, in China today there are more than sixty applications for genetically engineered animals, including several for fish . . .

A ban on GE fish in the U.S. would provide China with a global competitive advantage to launch similar products into international commerce."

This stuck in my craw.

"Do we really want to hold up China as this beacon of environmental regulation that we want to aspire to?" I said in response.

He barely took notice of my rebuttal. And as he continued, speaking of progress and opportunity and competition and the American way of innovation, his brushed-back gray hair practically bristling, Stotish morphed and blurred in my mind with the words and mannerisms of another chief executive with brushed-back gray hair. In my mind's eye I saw not AquaBounty's Ronald Stotish but the Pebble Partnership's John Shively. There was of course no direct connection between the two men. No business relationship. No means of profiting off the other's work. But the results of their separate endeavors could find a subtle coexistence.

If the EPA fails to make a 404(c) ruling and Pebble Mine goes through, the Pebble Partnership will profit enormously. But if an accident occurs, if an earthquake ruptures a retaining lake wall, for example, serious damage could be done to a system that produces two hundred million pounds of wild salmon a year.

If the FDA approves AquaBounty's fish for human consumption, the company could replace all of North America's unmodified farmed salmon with twice-as-fast-growing AquAdvantage fish, and the production of those farms could increase by two hundred million pounds.

There is a fearful symmetry in these two numbers. In the two hundred million pounds of potential wild loss and the two hundred million pounds of synthesized gain.

Artisanal Salmon

"How much time we got?" the Bristol Bay fisherman Reid Ten Kley asked me as he motored his skiff into position at the mouth of Alaska's Kvichak River.

"It's six fifty-one," I said. "You've got nine minutes."

"Shit," he muttered. Then, clipping his net line in place, he started hauling. Even as he made a little ground against the half ton of sockeye salmon that had rapidly filled the net on the ebb tide, the water exploded with more "hits" as fish newly arrived from the ocean smashed into the webbing. The haul became so ridiculous that Reid could only laugh, his voice full of the particular pleasure fishermen take in hard labor when the fishing is good.

"Here they come!" he shouted. "Heading to a sushi bar near you."

It was July 4. My birthday, incidentally. I had decided to give myself one last fishing trip to Alaska before I tried to sum up everything I'd learned about the fish fight of Bristol Bay. I had come on the invitation of Reid and his cousin Christopher Nicolson to a little spit of land called Grave-yard Point. Here, on the grounds of what had been the worst kind of industrial salmon cannery in the nineteenth century, Reid and Christopher and eighteen other men and women in their twenties and thirties had established a small-scale artisanal fishery for Bristol Bay sockeye salmon. Using their ancestral lease on the site (Christopher is part Athabas-can Indian and both he and Reid inherited their sites from their grandparents), they had begun marketing salmon under the name of the Alaskan great lake, Iliamna.

The Iliamna Fish Company's concept is simple, and it's something that is starting to spread around the United States. Instead of just giving their fish to a processor for a relatively low price and watching all that good fish disap-pear down the maw of global commodities trading, they would process it separately, brand it with the Iliamna name, and sell the fish directly to a small cluster of customers. This community-supported fishery (CSF) asks members in the cities of New York and Portland (Christopher's and Reid's off-season homes) to pay $220 a year up front for a share in its annual catch. This guarantees the fishermen something of an income, locks in the price against the vagaries of the market cycle, and provides the fishermen with some working capital at the traditionally cash-poor start of the season. In return, members get a monthly delivery of well-cared-for, carefully processed sockeye. A few select restaurants, like Bamboo Sushi in Portland, had also bought into the arrangement (this particular net haul was destined for Bamboo's sashimi platters). It was a good idea—one that provided Americans with a reasonably priced supply of healthy fish and Reid and Christopher with a decent income.

The only problem the Iliamna Fishing Company was facing right at this moment was that if Reid didn't get his nets hauled in by seven p.m., an Alaska Department of Fish and Game helicopter would drop out of the sky and issue him a four-thousand-dollar ticket. The continued success of this great Alaskan run is predicated on enduring clean water and specific, focused management tactics. Biologists carefully monitor every stream and close rivers constantly throughout the season to ensure that enough salmon of a range of different genetic varieties have made it upstream to spawn. This is something that irks some Alaskans. But they understand that someone needs to regulate the American catch today so that there can be an American catch tomorrow. It's a hard-won lesson, but one that is finally being learned.

Reid's section of the river was due to be closed at seven p.m. sharp, but the amount of salmon that had piled into his net was so great that he risked being out of compliance before the closure. At six fifty-two he and his crewmate Dusty doubled their efforts and leaned back against the gunnels, the twine cutting into their rubber gloves and stiffening their shoulders. In the flurry of activity I was sent out of the boat to retrieve fish that had washed up on shore with the receding tide. Encased in rubber overalls and lathered in sweat, I picked my way along the beach and gathered up salmon, putting a three-foot sockeye on each of my available fingers until I looked like some kind of perverse Christmas tree—literally covered in salmon. Finally, when Reid shouted at me and tapped his watch, I sloshed back, my arms weighed down, live salmon in the river bumping against my knees. I hopped aboard just as Reid finished the pull with a single minute to spare.

Writers like me, trying to convey something about the state of the natural world, find ourselves in a tricky spot these days. There is a lot of bad news out there, but we know that readers do not want to hear any more about it. They are tired of global warming, acid rain, the loss of biodiversity, overpopulation, and the like. And so when we fashion our narratives it is with Pollyanna on our shoulders. No doubt the reader can detect that there is a dark void beneath our feet in spite of the positive stories we try to tell. But at the risk of sounding like a consummate Pollyanna, I found as I sat at the head of Reid's boat, the wind blowing back my hair and a rainbow arcing ridiculously out across the sky, that I was filled with hope for Bristol Bay. It seemed things were looking up. People were starting to change their minds about just how important the bay was for the country.

It's hard to say what had tipped the balance. It may have been Rick Halford's private aerial tours of the Pebble Mine deposit, offered to anyone who would listen. It could have been the Yup'ik BBNC board member Pete Andrew shooting his little goodwill jars of smoked salmon across the tables of power in Washington, D.C. It could have been Justice Sandra Day O'Connor and her bull-moose stand for fish and nature. Or maybe it was the spirit of the land itself. Whatever the cause, the creaking levers of power had started to shift in Bristol Bay's favor. In May 2012 the EPA, against the vehement protestations of the Pebble Partnership, published a first review draft assessment of Bristol Bay, concluding, "Though precise estimates of the probabilities of failure occurrence cannot be made, evidence from the long-term operation of similar large mines suggests that, over the life span of a large mine, at least one or more accidents or failures could occur, potentially resulting in immediate, severe impacts on salmon and detrimental, long-term impacts on salmon habitat." The Pebble Partnership called the assessment "dangerously rushed," but nevertheless it was released to the public and Americans outside Alaska had a chance to weigh in on it. For the next year there would be hearing upon hearing, conferences and debates around the country to determine the fate of the great salmon grounds of Bristol Bay. These hearings would be packed with a majority of Pebble opponents. Pebble would fight on and on and delay on and on. It would protest the first watershed assessment in 2012 and call for a scientific review. The review would be published with even more vehemence about the inadvisability of the mine. In January 2014, a second watershed assessment would be published by the EPA, with, again, strong words of warning for mining in this sensitive region. Senator Mark Begich of Alaska would at last, after this second assessment, come out firmly against Pebble Mine. And finally, in Feburary 2014, the EPA would take the first regulatory steps toward initiating a 404(c) Clean Water Act process in the Bristol Bay watershed—something that could protect the fishery for the indefinite future.

The Pebble Partnership, though, has suffered a major blow. Amid all the political finagling, Anglo American, the

British mining giant, decided to withdraw from the Pebble project and take a $300 million tax write-off for the loss incurred developing the proposal. The remaining partner, Northern Dynasty, has vowed to continue. Indeed, why shouldn't it? Anglo American has already funded over $500 million in research and development, arguably one of the most expensive parts of bringing a mine to market, and who knows what winds will blow Pebble's way should a different administration occupy the White House? A $500 *billion* copper and gold deposit is worth waiting for.

The fight will drag on. Nonprofits and activists will come and go. Writers will write. Opinionators will opine. But as the future of this fishery continues to teeter, I am certain there is one group of people who will keep their dogs in this fight: the people who cannot afford to lose this battle, the fishermen of Bristol Bay.

It is the general position of the environmental community to see fishermen as the ocean's bane—rapers of the environment, stealers of nature's bounty. This certainly cannot be denied in many instances. But in the case of Pebble Mine it is fishermen who have come to the defense of the fish. It was fishermen who, first and foremost, noticed that something was amiss. It was fishermen who first traveled to Washington to raise the issue with our elected officials. And for the first time perhaps in history all kinds of fishermen united. Commercial set netters and gill netters. Fly fishermen who normally eschew such company, seeing commercial fishermen as thieves of their sport. Fathers and their children who fish to put meat on their daily table and rejoice in the state's subsistence fish allotment. They were united for once in a common purpose: to protect the sockeye. They understood what others did not: that nature will reward us—will literally feed us—if we take care of it.

It was about eight p.m. when we finally got our four thousand pounds of sockeye salmon over to a tender vessel that would in turn bring it to a processor, where it would be filleted, packaged with the Iliamna label, and sent to the members of the Iliamna community-supported fishery in New York and Portland. Fish dispatched, our skiffs convened and motored together back to Graveyard Point. There was some confusion as to when the next open fishing period was going to happen. Some thought it was one thirty a.m. Others said it was one thirty p.m. But we wouldn't know until Fish and Game made a final announcement around ten that evening. Actually, it didn't matter all that much since the sun barely sets in the boreal summer and all anyone really wanted to do up here was fish. Once we were back at Graveyard Point, most of the fishermen of Iliamna disappeared along the raised planks that led into a maze of bombed-out old cannery buildings to try to grab an hour or two of sleep. The whole thing felt like an odd mix of a Jack London novel and a college campus in the midst of an all-night cram session.

Not feeling sleepy, I ducked into the most solid-looking building, a corrugated-tin affair labeled "The Fish House." Inside I found the camp cook, a petite woman in her thirties named Cricket, in the midst of dinner preparations. She offered me a choice of beverages: coffee, tea, or Tang. Alaska is the great repository of the foods that time forgot. Along with the Tang were Pilot hardtack biscuits, Smucker's grape jelly in an oversized old-style jar, and I thought I spied a container of Ovaltine in the pantry.

But Cricket did yeoman's work with the materials at hand and got the big gas range into a lather. Soon the room was filled with the good aromas of hearty food cooking in weighty iron skillets that little Cricket deftly shuttled across the range. And when the smell of her cooking drifted out over the fish camp, it drew the drowsy fishermen in for the evening meal. After all the men and women had gathered, I was surprised to see Reid, the ultimate hipster-seeming Portlander, bow his head in a most earnest prayer.

"Heavenly Father," Reid said, "thanks so much for the

really good fishing today. I mean just *really* good fishing. Thanks for the calm weather that let us go fishing. Thanks for the good food that Cricket made." In its simple way the prayer touched me and made my usually God-free heart skip a beat. As Reid's cousin Christopher put it to me, "I think the sea is God's answer to agnostics."

The same, I thought, can be said of everything that lives in the sea that nourishes us. To eat from dry land, we humans are compelled to tame the countryside, till and fertilize it, kill off pests and fence off territory. To grind and process and flavor. Add nutrients. Subtract fat. Shape and mold until that increasingly artificial landfood becomes something not quite food.

At sea everything is a miracle by comparison. All that the sea asks of us is that we be wise in our harvest, recognize the limits of its bounty, and protect the places where seafood wealth is born. In return the sea will feed us and make us smarter, healthier, and more resilient. Quite a covenant.

With dinner stowed away I took a walk around the old cannery grounds of Graveyard Point while the polar sun circled in the nightless summer sky. This was a place where industrial fishing had once turned a dynamic wild animal into a faceless commodity product that traveled the world in a can. I thought of the particularly contemporary American reworking of those bad times that was happening here today—a small group of committed artisan fishermen doing something very different. A group that fished modestly, followed the rules, caught what could be sustained, and formed their own pact with American fish eaters to the south—fish eaters who wanted to know what they were eating, how it was caught, and how the people who caught it led their lives.

Passing up to a bluff, I looked down on the isolated little settlement washed by the foam of the Bering Sea and thought that, once upon a time, a little seventeenth-century fishing village called New Amsterdam must have looked quite a bit like this. A modest place with its face turned toward the sea where the fisherman and the fishmonger were an integral part of daily life and where seafood held its own with landfood in nearly every regard. What kind of alternative future might America have had if the descendants of New Amsterdam had decided that seafood, not landfood, would be the new country's bread and butter? Not a Jeffersonian society that relied upon the harvests of agriculture, nor a Hamiltonian one that captained banking and industry, but rather a "Neptunian" democracy that lived off the sea. A place where estuaries were recognized as the heart of the food system, where rivers were rarely fettered for fear of impeding the miraculous runs of salmon, sturgeon, shad, and herring, where human development moved in parallel with the protection of the near shore. What kind of society might we have formed had we not, as Melville wrote in *Moby-Dick*, become "landsmen . . . tied to counters, nailed to benches, clinched to desks." What if instead we had become what Melville called a society "fixed in ocean reveries"?

My own reverie was interrupted when I heard a bustling in the encampment and saw Iliamna's fishermen emerging from the old cannery buildings. One by one they headed to the Fish House. Word had gone out that an announcement was about to come over the radio on the status of the run, and no one wanted to be late to their boats in case an opening was declared. When everyone had gathered, Reid dialed the shortwave through the static until finally locating a chipper, almost Midwestern-sounding Alaskan voice. "This is the Alaska Fish and Game Department," the voice declared perkily in spite of the late hour. Brightly the announcer told how the previous night's wind and rain had set the salmon moving. In the last six hours an additional million fish had passed by the department's observers and escaped into the upper river—far more fish than had been expected. "Due to the surge in fish passage," the voice concluded, "Fish and Game has determined that additional fishing is warranted. The next opening in the Kvichak region will be from one thirty a.m. until ten thirty a.m." A wave equal parts exhaustion and excitement rippled over the assembly. And then all made a beeline for the mudroom, where their overalls hung. Everybody would need to bolt their coffee and rush to get ready in time.

It was time to go fishing.

Conclusion

We are what we eat, we are told. But we Americans do not eat what we truly are. We are an ocean nation, a country that controls more sea than land and more fishing grounds than any other nation on earth. And yet we have systematically reengineered our landscapes, our economy, and our society away from the sea's influence. As of 2012, Americans ate a little less than 15 pounds of seafood per person per year, well below half the global per capita average and minuscule in comparison with the 202 pounds of red meat and poultry we consume.

Lately we have started to understand the health benefits of seafood and this is in part why seafood consumption in the United States has risen somewhat in the last forty years. But now that we are beginning to realize that our diet should be much fishier than it is, we are finding that our biological and economic seafood infrastructure has been hollowed out. When thirteen of the fifteen-odd pounds of seafood we do manage to eat is foreign in origin, it speaks to a glaring problem of nutritional food insecurity, one that derives from our decision over these last centuries to become a progressively less and less fishy place.

We became an unfishy place and developed landfood to the detriment of seafood in a very understandable way. When our colonial forebears arrived here, they set out to emulate the agricultural societies from which they hailed. They filled in the near shore, diked the natural ebb and flow of land and sea wherever they could, drained away marshland at every opportunity, dammed rivers right and left, and made America's naturally seafood-rich waters poorer. After so much of the near shore started to lose its seafood-producing potential and the memory of our coasts as food systems started to fade, we proceeded apace. We built up industry at the expense of clean water, developed commodity agriculture to the impairment of locally diverse marine food systems, and opened up avenues of global trade that led to the devaluation of our home-produced fish and shellfish. We still caught fish but mostly we caught them offshore en masse using fishing technology that tore up off-shore seabeds and further degraded the state of American fisheries.

We did all of this more or less unconsciously. We did not

know that when we drained our salt marshes and turned them into fields for soy and corn and other commodity crops we were destroying the spawning and rearing grounds of 70 percent of our wild seafood. We didn't realize that by removing our natural oyster beds we were cutting the fish-carrying capacity of our estuaries in half. We failed to notice that when we threw up over eighty thousand large dams and millions of other smaller obstructions on rivers and streams all over the country we were eradicating the runs of salmon, sturgeon, shad, and herring that provided real food for the nation. We were in denial that when we dumped trillions of gallons of sewage and industrial effluent into our waters we were lacing our seafood with harmful bacteria and persistent organic pollutants and heavy metals. And we didn't quite believe that if we increased our coastal population, as we have by a whopping 39 percent over the course of the last forty years, we would overrun working fishing communities, degrade water quality, and hobble our coasts' ability to feed us.

Now we know.

But to date our response to this knowledge has been mechanical and shortsighted.

Fear of pollution, greed-driven real estate development in our estuaries, and worry about domestic overfishing have compelled us to try to fix our seafood shortcomings through foreign imports. Instead of confronting the essential biological infrastructural problems of our domestic seafood supply, we have decided to rely on multinational corporations and cheap foreign production to meet our needs. And while the locavore American food movement has spurred the production and consumption of local landfood beyond all expectations, local seafood remains mostly a quaint curiosity. We are too confused or too preoccupied to widen the circle of localness to include the local ocean.

All this in spite of the fact that improved American fisheries management has indeed worked. As we approach the twentieth anniversary of the Sustainable Fisheries Act biologists are finding that two decades of careful regulation have allowed American fish to gradually recover from the lows of the 1980s. Overfishing has greatly decreased in American waters. But American *fishermen* are going extinct. Having lost their market share to the bargain-basement prices offered by foreign aquaculture, American fishermen can no longer compete. This in turn causes more fishermen and shoreside fish-processing centers to go out of business. The lack of processing capacity means that fishermen are increasingly selling whole unprocessed fish to foreign companies that gladly take it, freeze it, ship it, defrost it, process it, and market it to customers in their home nations. It's for this reason that things like New Bedford scallops, California spiny lobster, Alaska salmon, and Oregon black cod mostly end up getting sold abroad.

Meanwhile, with fishermen and processors no longer occupying the coast, developers can come in more easily, privatize the waterfront, and drain the wetlands—the very nurseries that make wild fish and shellfish possible. The new residents of these seaside retreats gladly buy foreign-farmed shrimp at their local supermarkets and enjoy their shrimp cocktails obliviously as the sun sets over their verandas.

It's not necessarily the aquacultured aspect of the shrimp on those verandas that I object to. I have in the past advocated for marine farming, even found its essential premises quite tenable in the macroeconomic picture. In many cases aquaculture *is* more efficient than fishing. It *can*, in many instances, cause less environmental harm than fishing. But the major harm aquaculture does is something invisible and ideological that the average American consumer just doesn't notice.

Plainly and simply, aquaculture makes wild estuaries and marshes seem irrelevant. Innovators in the 1960s and 1970s took apart all the diverse elements of wild estuaries and reassembled them in a laboratory, decoupling the wild environment from the consumer. When every last piece of seafood you eat has been grown on a farm, what need do we have for marshes, estuaries, and wetlands of all kinds?

Very little, it turns out. This is reflected in the top six seafoods the average American consumes:

1. Shrimp: 90 percent imported, mostly farmed in Asia
2. Canned tuna: mostly imported and harvested far offshore in waters that are not reliant on coastal estuaries
3. Salmon: two-thirds farmed, almost entirely from abroad
4. Alaska pollock: wild and American, but, again, harvested far offshore, mostly unrelated to the health of the coast
5. Tilapia: farmed and imported, often from China
6. Pangasius catfish: farmed and imported entirely from Asia

These top six seafoods now to a large degree comprise that 91 percent of the ocean products that Americans eat. The remaining 9 percent—the seafood that is reliant on the health of the American coast—is pocket change.

All of this has happened mostly out of sight of the average consumer. For hand in glove with the shift in origin of our seafood over these last fifty years has come a shift in the place where we buy our seafood. Once upon a time a visit to the local fish market was something Americans everywhere did in the course of their week. Such a visit gave a window into the health of the local ocean and in its own way educated the consumer about which fish were thriving and which were faring poorly. It was also a way for Americans to come in contact with people who knew the business of the ocean. But from 1985 to 2005 fish markets and individual fishmongers went from controlling 60 percent of the U.S. seafood trade to holding on to just 10 percent. Supermarkets, meanwhile, went from selling around 16 percent of our seafood to selling more than 80 percent. Is it any wonder that industrially processed foreign product has replaced the familiar faces of the fish next door?

Subtly, almost clandestinely, though, a reaction against the commodification of our seafood supply is starting to emerge. I witnessed this on a crisp New York autumn morning in 2013 as I walked through the neighborhood of lower Manhattan that I call home. I turned down Broadway, as I had eight years before, now knowing after Hurricane Sandy that Broadway is a ridgeline and that the low-lying shores on either side of the island of Manhattan can flood on a full moon and a high wind. I headed downhill toward what had been the center of New Amsterdam, a place where colonists traded in all sorts of wild creatures but which is now a financial district that deals mostly in stock options, derivatives, and other things that have no real value in the world. And as before I turned east down De Maagde Paatje, I passed over Parelstraat, and at last I came to Fulton. Dear old Fulton, that still boarded-up question mark of a market that occupies the East River esplanade beneath the Brooklyn Bridge, a place that in its silence poses an open-ended question about what we should do with our waters, our fisheries, our last truly American food.

But today it is not silent. Today is market day. Today the New Amsterdam Market, a small group of local citizens, has decided to hold a "gathering of fisheries" in the parking lot of the old Fulton Fish Market. For the last eight years the New Amsterdam Market has been trying to rescue the old Fulton buildings from real estate development, and now it has invited a dozen-odd fishermen, community-supported fisheries, smoked-fish vendors, and oyster farmers to present their wares directly to customers. The two rows of mongers face each other, some puzzled, some excited, all of them wondering out loud whether this couldn't somehow be a different model to make Americans sit up and take notice of their marine environment.

The New Amsterdam Market was founded by a somber

and determined Italian American named Robert LaValva. The great-grandchild of Sicilian fishermen, LaValva had been studying the old market district for years and had decided that it should not simply slip into oblivion. He was faced with a formidable opponent in this endeavor. All around, the Dallas-based Howard Hughes Development Corporation was pressing in, building a new South Street Seaport mall, and making moves to redevelop the entire district. Of late, the Hughes firm had been making noises about demolishing some of the old market, including the "New Market Building," to make room for a luxury fifty-story hotel and residential tower. But LaValva had fought back relentlessly. Again and again he made the point that the old Fulton Market sat on publicly owned waterfront and that no one had the right to privatize it. For eight years he staged special markets—an oyster saloon that recalled the great New York oyster trade of years gone by, a cheese and dairy fair, a congregation of New York hard cider makers.

And now he had made a fish market, a place for people who have a stake in making the waterfront alive and viable to come together again.

When I reach the market, I see J. P. Vellotti, an oyster producer from Westport, Connecticut, the place nearest to New York City where viable oysters are still grown. He's selling oysters as well as shucking knives that he's fashioned from the old mahogany deck boards of one of the last big oyster sloops on Long Island Sound. He tells me he's just received permission to grow scallops in addition to oysters and is hoping to launch his product sometime next year. I think about the shellfish beds that border our towns and cities where we have the technical capacity to revive shellfish farming again. We have made tremendous strides in the last forty years in bringing back oysters, figuring out their life cycles, and using the Clean Water Act to create an environmental space for their sustenance in the waterways we've improved. We have in fact doubled oyster production in the last forty years. But this is just a start. Even with this doubling we are still at only 14 percent of our historical oyster production capacity. In addition, there is a range of other creatures we could bring into much larger-scale commercial culture. Clams, mussels, scallops, kelp—all of these creatures that clean and filter tremendous amounts of water— double the quantity of commercial finfish species by the habitat they provide and give us food that is high in essential nutrients. All of that could become the cornerstone of a new coastal economy. As a Maine-based oyster and mussel grower put it to me once, "Shellfish are an economic argument for clean water." With a coastal economy whose foundation is based on this very argument, we could create a positive-feedback loop that links clean water to healthier food and a more productive marine environment.

Near J.P. I run into Andy Willner, the founder of NY/ NJ Baykeeper and the man who first tried to reintroduce oysters to New York Harbor back in the 1990s. We talk about how, as the water continues to grow cleaner around New York, more and more oysters are turning up. How underneath the Tappan Zee Bridge so many oysters were found in the summer of 2013 that the Army Corps of Engineers is flummoxed as to how to renovate the bridge without disturbing the reemerging beds. I think about all the excitement going on around the waterfront in the wake of Hurricane Sandy. Kate Orff, the oyster-tect, has just advanced to the next phase in a region-wide design competition and produced a vision of the New York Bight called the Shallows—a vision where educational institutions are linked to all the watersheds that once had oysters in the greater New York Bight. The Harbor School is furnishing the educational tools to introduce an oyster curriculum at all those schools.

Whether we choose to embrace the ocean or not, it is coming to embrace us, faster than many of us can believe. Two feet of sea-level rise in the next fifty years in New York City alone is a change that will force us to literally redefine the American waterfront and how we interact with it. I think about the danger and an opportunity in this. If history is any indicator of what our reaction will be, the first, knee-jerk response to the coming inundation will be to try to keep the water out. To harden our shorelines even more than we've done already. To reject the biological in favor of the mechanical. But at best this would be a short-term solution. Barriers break. Salt water is pervasive and corrosive. There is no real stopping the sea. And so why stop it? Perhaps the answer lies in a conjoining of the interests of seafood and the interests of humans. Both would benefit from a soft-edge approach.

Just up from the oyster stand, I run into Jason Delacruz, a Gulf coast red snapper and grouper fisherman and monger who's come up from Florida. On ice Jason has a dozen whole fish, each individually labeled with a "Gulf Wild" tag, a system that allows each and every fish he sells to be traced back to the captain who caught it. Originally started as a response to the 2010 Deepwater Horizon oil spill in order to reassure customers that the fish they were eating hadn't been caught in polluted waters, Gulf Wild has expanded into a much larger concept of trying to put a face back onto the fish we eat. "The old buyers, they didn't *want* the fish to be identifiable, they didn't *want* the fisherman in the system," Jason tells me. "It was better for them if they could just move it as a commodity." But with Gulf Wild, Jason's fish are identifiable from boat to plate. Fewer middlemen get involved in the process. More value goes from the consumer to the fisherman and as a result greater attention can be paid to sustainable harvest and careful handling of the product. I ask Jason if they are thinking of trying to include shrimp in the Gulf Wild brand. "You know, it's funny you should mention that," he says with a twinkle in his eye, realizing the sheer magnitude of the shrimp trade and the value that could be preserved if shrimpers had better control over their product. "We want to get that going. We're looking into the sustainability aspect of that. How to determine if shrimpers are using their turtle excluders correctly. How to bring that product to market." I offer to put him in touch with Tom Hymel and the people doing DelcambreDirect in Louisiana. There are many uncertainties, but the desire for change—for a new system—is there.

Over to my right are the Bristol Bay salmon fisherman Christopher Nicolson and his wife, Emily, now back in New York after the season's sockeye fishing has concluded in Alaska. Their homeschooled kids are at Presbyterian catechism today and technically Nicolson should be with them. Indeed, he seems a little bit ashamed to be selling fish on the Lord's Day. "It's just terrible," he jokes, "making mammon on a Sunday!" His big, strong hands parse out great slabs of a new smoked-sockeye product he's been developing with a smoker in nearby Mount Kisco, while Emily takes down names of people eager to sign up for shares in next year's community-supported fishery. We talk about the Bristol Bay campaign, chat about Anglo American pulling out of the Pebble Partnership, and between pats on the shoulders to his regular CSF customers muse about what could happen in the future with Bristol Bay. We talk about the myriad pressures that threaten to gut Alaska's seafood potential just as that potential has been gutted in the lower forty-eight, and we wonder what it will take for the United States to finally realize the paramount importance of the seafood-producing aspect of America's fiftieth state. We have a strategic grain reserve in this country and a strategic oil reserve. Why couldn't Alaska be our nation's strategic seafood reserve?

As I bid good-bye to Christopher, I think about what it would be like for his CSF method to become common practice for American fisheries. Imagine a seafood system in which every fishery is linked to a community that buys into its potential every season, a system in which the community's fish supply is directly dependent on the health of the local waterways. Across from us, Sean Barrett, from a new Montauk-based venture called Dock to Dish, would surely agree. A fisherman and restaurateur, he has just started a business that connects New York restaurants directly to Long Island fishermen, preserving the value chain for fishermen and ensuring that what is caught is well cared for. Indeed, Nicolson and Barrett are just two examples of something that could almost be called a trend. In the last five years at least forty different community-supported fisheries have been born and are now active in nearly every American coastal state. But there's still a long way to go.

And looking over the sweep of them all—all the fishermen, the mongers, and the throngs and throngs of people who are now crowding the market and buying up all the fish (everyone will sell out that day, I learn later)—I wonder about whether in the United States something could be made from this alliance I see forming before my eyes between fish catcher and fish buyer. Whether we couldn't hope to develop a seafood sector that once and for all breaks away politically and culturally from other extractive enterprises—the oil and gas industry, commodity agriculture, timber, real estate development—and begin to recognize that fishing has real financial interests that lie in direct opposition to these more harmful ways of doing business with the planet.

As a group, fishing people trend to the conservative side of things. They tend to resist government regulation and turn a jaundiced eye to people who call themselves environmentalists. But as our coasts get more crowded and contested, fishermen will have to choose a pathway that diverges from this mentality. Their real points of contention are of a different nature. The real fight ahead will be over the health and integrity of the ecosystems upon which their livelihood subsists. This is, in the end, a fight for all Americans. A struggle for biologically vital coasts, economically viable waterfront communities, and good, healthful food.

Acknowledgments

Many people with long memories of the past and active participation in the present of American seafood contributed greatly to this book.

In the realm of the lost Northeastern oyster kingdom, I am deeply grateful for the advice of CUNY's John Waldman, Jim Lodge and Clay Hiles at the Hudson River Foundation, Mark Kurlansky, the Malinowskis (Steve, Sarah, and Pete in particular), the Thimble Island Oyster Company's Bren Smith, Cathy Drew, Taylor Shellfish Farms, NY/NJ Baykeeper, Jon Rowley, the Milford Laboratory at NOAA's Northeast Fisheries Science Center, the Urban Assembly New York Harbor School, Katie Mosher-Smith, and all the growers trying to get oysters going again. With an eye toward the future, a deep bow and thanks to Kate Orff and the team at SCAPE who are trying to imagine a fishier New York. And thanks, too, to my Opinion Page editor at the *New York Times*, Clay Rosen, for being willing to listen to the soft whisper of the oyster's voice.

On the Gulf coast, a huge debt of gratitude to the writer, radio host, and Gulf seafood champion Poppy Tooker, who helped me break through the press-weary haze postspill and speak to people who feel passionately about the continental United States' most important fishing grounds. The many tendrils leading out from Poppy brought me to other extremely helpful fish folk, including Jim Gossen of Louisiana Foods, Richard Mc-Carthy, formerly of the New Orleans greenmarket, Albert "Rusty" Gaudé III of Louisiana Sea Grant, and Al and Sal Sunseri of the P&J Oyster Company. On the scientific side of things in the delta, much appreciation and affection to Paul and Libby Hartfield (Paul of the U.S. Fish and Wildlife Service, Libby of the Mississippi Museum of Natural Science). A similarly deep thanks to the Gulf dead zone researchers Nancy Rabalais and R. Eugene Turner. Also thanks to Olivia Watkins and Chuck Perrodin, both of whom helped me liaise with the state of Louisiana's politicians, scientists, and regulators.

A general word of thanks to the state of Alaska for showing me what intact fisheries really should look like. Within that great state Pete Andrew, Jason Metrokin, and the Bristol Bay Native Corporation deserve particular praise for their help and for their determination in the Pebble Mine fight. Also to Bob Waldrop of the Bristol Bay Regional Seafood Development Association, Rick Halford, Elizabeth Herendeen, Beth Poole, and of course Shoren Brown, even if he does need to work on his back cast a bit. And thanks to Christopher Nicolson and the rest of the crew at Graveyard Point for keeping the nets full and the bears at bay.

Once all of this had been put on paper in raw form, I relied on many superb writers and editors for comments and help. Among them were Rowan Jacobsen, Cressida Leyshon, Tara Bray Smith, John Donohue, Sarah Schenck, Sean Wilsey, Daphne Beal, Barry Estabrook, Clare Leschin-Hoar, Katherine Baldwin-Eng, Molly Shapiro, and Mollie Boero. Chief of all constructive critics was the contractor herself, Lindsay Whalen, my editor at Penguin. I would have stopped at three drafts. Lindsay took me to seven and the work benefited greatly from her yoke. Through those many drafts deep emotional and creative support was provided by my literary agent, David McCormick. A writer could ask for no better representation. Relief pitcher of the year goes to the historical marine ecologist Carolyn Hall for her ninth inning research and fact-checking at the conclusion of the project.

For moral and material support, a big thanks to Carl Safina and the Blue Ocean Institute. I know of no other person who feels the sea as well as Carl, and I am deeply in his debt for help on every aspect of this project. Also many thanks to the Walton Family Foundation for supporting my research work in the Gulf, and to the Food and Environment Reporting Network, whose funding helped me understand the unfortunate influence big agriculture continues to have on small seafood. And many, many thanks to Guy Lancaster, the Still Point Fund, and the South Street Seaport Museum for supporting me through a writer-in-residence position during the important final months of this book's completion.

In that same arena, thanks to Robert LaValva, the man who has for eight years worked tirelessly to keep the old Fulton Fish Market from being dumped into the sea. I must take off my fishing hat and wrap a symbolic Saint Jude medal around his neck. It's still too early to say whether Fulton is a lost cause or not, but Robert has been fighting an unpaid uphill battle for authenticity over commodification all these years and deserves all the medals real or imagined that can be minted.

Finally, thanks to Esther, Harvey, Sharon, Nini, Helen, Bud, Phil, Tanya, and Luke for being forces for good at all times. A person feels less underwater when supported by so many buoys.

Notes

1   Heart disease, Alzheimer's, depression, even low sperm count are all conditions that a fish-based diet may help ameliorate: Specifically it is the omega-3 fatty acids in fish oil that potentially reduce cholesterol and heart disease risk. "Omega-3 fatty acids may decrease triglycerides, lower blood pressure, reduce blood clotting, boost immunity and improve arthritis symptoms . . ." From Mayo Clinic studies: www.mayoclinic.com/health/omega-3/HB00087. Also from the Mayo Clinic, low levels of EPA and DHA are associated with depression so there is some evidence that fish oil can help alleviate mild depression. www.mayoclinic.com/health/fish-oil-supplements/AN01399. Regarding Alzheimer's, a 2011 ten-year study presented at the annual meeting of the Radiological Society of North America compared fish consumption and retention of gray matter in the brain. The findings showed that weekly fish consumption was positively associated with gray matter volumes of the brain that are strongly linked with Alzheimer's, thereby reducing risk in those areas. From ScienceDaily.com: www.sciencedaily.com/releases/2011/11/111130095257.htm. And as for low sperm count, the jury is still out. But because omega-3s lower cholesterol and boost DHA, which is also found in the testes, there are some studies that associate fish oil with sperm health—more along the lines of healthier sperm rather than more sperm, but possibly more effective for fertility. Also see Akhlaq A. Farooqui, *Beneficial Effects of Fish Oil on Human Brain* (Philadelphia, PA: Springer Science + Business Media, LLC, 2009).

1   **The United States controls more ocean:** The United States has full sovereignty over waters that extend from the low tide line out to twelve nautical miles. From twelve nautical miles out to two hundred nautical miles is what is known as the country's Exclusive Eco-

nomic Zone (EEZ). EEZs are not technically the property of the adjacent nation, but according to contemporary maritime law a country has exclusive rights to the marine resources beneath the surface of EEZ waters. Peru, Ecuador, and Chile implemented EEZs as early as 1952. EEZs were widely implemented in the 1970s and 1980s after the United States unilaterally did so through the passage of the Magnuson-Stevens Act. With coasts on three sides plus numerous island holdings as well as Alaska, it is understandable how the United States has the world's largest EEZ. France, by the way, is number two. A map of U.S. EEZs can be found at http://www.gc.noaa.gov/documents/2011/012711_gcil_maritime_eez_map.pdf.

1 **91 percent of the seafood Americans eat:** The so-called seafood deficit is not a trade balance number in dollars. As Fionna Matheson at the National Oceanic and Atmospheric Administration (NOAA) wrote me, the seafood deficit "is measured in pounds, and comes from this formula: Domestic production + imports − exports = total consumption. Total consumption is divided by the total U.S. population to generate an estimate of per capita consumption." The figure of 91 percent is indeed a striking amount. That said, Matheson and others qualify it, calling it a somewhat "squishy number" largely because a certain and unknowable portion of foreign seafood is seafood America exports to Asia, where it is processed and then sold back to us in a can or in a frozen package. This phenomenon is detailed in the salmon chapter of this book. As quoted in NOAA's 2012 summary of seafood trade: "Exports may include merchandise of both domestic and foreign origin. Census defines exports of 'domestic' merchandise to include commodities, which are grown, produced, or manufactured in the United States. For statistical purposes, domestic exports also include commodities of foreign origin which have been changed in the U.S. from the form in which they were imported, or which have been enhanced in value by further manufacture in the United States. . . . For example, fish sticks cut from imported fish blocks, when exported to a foreign country, are classified as domestic for statistical purposes." This means "domestic" does not relate to the home waters where the fish were caught. Some percent of our imports are actually from U.S. waters whereas some percent of our exports are from foreign waters. Do these percentages cancel themselves out? We don't know. NOAA data can be obtained at www.st.nmfs.noaa.gov/Assets/commercial/trade/TRADE2012

.pdf. It should be noted that other sources, such as the Alaska Seafood Marketing Institute, put the amount of foreign seafood consumed by Americans at closer to 80 percent of our fish diet.

2 **five times the national per capita rate of seafood consumption:** In 2012 the United States produced 25.91 billion pounds of beef. The U.S. 2012 population was 313.9 million people. That means there were eighty pounds of beef produced per person in the United States—and the 2012 U.S. consumption of beef was 25.8 billion, nearly the total production (www.ers.usda.gov/topics/animal -products/cattle-beef/statistics-information.aspx#.Uop-uhboc0Y). The seafood per capita consumption was 14.4 pounds per person in 2012, a drop from 15 pounds per person in 2011. www.st.nmfs .noaa.gov/Assets/commercial/fus/fus12/FUS_2012_factsheet.pdf.

2 **In the last half century American seafood imports have increased by a staggering 1,476 percent:** Calculated percent increase in edible export volume (metric tons) from 1937 to 2012. Those wishing to delve into the import and export data of the last fifteen years can visit NOAA's fisheries statistics site at www.st.nmfs.noaa.gov/ commercial-fisheries/foreign-trade/index.

2 **a third of the seafood Americans** *catch* **gets sold to foreigners:** This is in terms of edible seafood. In 2012 the total U.S. catch was 9.6 billion pounds; the edible exports were 1,425,591 metric tons, or 3.14 billion pounds; so the export was a third of the catch.

2 **every year, being sent abroad, more and more often to Asia:** A search through exports to individual countries shows our exports to China of salmon, groundfish, and flatfish growing exponentially over the last twenty years compared to most others. The 2012 figures for fisheries trade can be found here: www.st.nmfs.noaa .gov/Assets/commercial/trade/TRADE2012.pdf and import or export by country found here: www.st.nmfs.noaa.gov/commercial -fisheries/foreign-trade/applications/annual-product-by -summarized-countryassociation.

3 **bare asphalt slab called Zuccotti Park:** Many will recognize Zuccotti Park as the site of the 2012 Occupy Wall Street protests. In my home, dinners in the fall of 2012 were usually accompanied by cheers or boos that came through our windows from the Occupy crowds.

4 **had been called De Maagde Paatje:** Background information on the geography of the New Amsterdam colony is drawn primarily from Russell Shorto's excellent *The Island at the Center of the World* (New York: Doubleday, 2004).

4   **a red clapboard building:** The tavern at the corner is now known as the Bridge Cafe. It was originally a brothel when it was constructed in 1794, and was converted into a bar in 1847. It is rumored to be haunted by a female ghost. As of this writing the Bridge Cafe is still closed owing to damage sustained by Hurricane Sandy.

5   **Fulton Fish Market:** The remnant fish market comprises two buildings, the Tin Building, which sits just to the back of Pier 17 on the East River, and the adjacent New Market Building, which is described here. The New Market Building was built in 1939 and was considered a state-of-the-art facility in its day. A complete report on the building and arguments for its preservation are contained in a 2009 report of the Municipal Art Society of New York available online at mas.org/wp-content/uploads/2009/02/New -Market-Building-Report.pdf.

6   **not as distant as one might think:** A useful and concise description of Fulton and its formation is contained in John H. Matthews, "History of the Fishery Industries of New York," *The Fishing Gazette*, 1929, 49–80.

8   **halt Westway in its tracks:** An account of the Westway affair can be found on pages 131–137 in John Waldman's *Heartbeats in the Muck*, revised edition (New York: Fordham University Press, 2013).

9   **fishmongers were required to sign a new market lease:** There was a provision in the New Fulton Fish Market Cooperative Lease that stated that vendors agreeing to move to the new facilities in Hunts Point would never sell again at the Old Fulton Fish Market in Manhattan. It is Exhibit F: Form of surrender and termination agreement. Lease obtained through a request to the New York City Economic Development Corporation pursuant to the Freedom of Information Law ("FOIL"), Article 6 of the Public Officers Law.

9   **Fulton's departure can be seen as a sign:** From page 120 in J. L. Anderson, F. Asche, and S. Tveterås, chapter 8: "World Fish Markets," in *Handbook of Marine Fisheries Conservation and Management* (New York: Oxford University Press, 2010).

9   **Supermarkets, meanwhile, went from selling 16 percent of our seafood to selling 86 percent:** Ibid.

9   **From 1985 to 2005, the same period during which our seafood imports doubled, our seafood *exports* more than quadrupled:** Ed-

ible seafood product import and export. www.st.nmfs.noaa.gov/
Assets/commercial/trade/TRADE2012.pdf.

10 A nation where nearly half the population chooses to live less than
ten miles from the sea: In 2010, 39 percent of the U.S. population
lived at or near the shoreline. That amount is projected to increase to 47 percent by 2020. http://oceanservice.noaa.gov/facts/
population.html.

11 a place where local oysters were associated with the worst kinds of
food-borne illnesses: In 1880 the New York City population was
1,206,299 people and local oyster beds were producing 700 million oysters a year. Oysters were associated with typhoid and
blamed for cholera, and by the 1920s oyster harvesting in New
York was being shut down. From the U.S. Census and from pages
244 and 262–263 in Mark Kurlansky, *The Big Oyster* (New York:
Viking, 2006).

11 entire oyster industry hovers at a mere 14 percent of what it had
been at its peak: According to Robert Rheault of the East Coast
Shellfish Growers Association, all the historically producing
estuaries and bays on the Atlantic are at 5 percent or less of peak
production—with the Hudson, Delaware, Chesapeake, and Narragansett all at 0–1 percent. National production includes the Gulf
and Pacific coasts. In the early 1900s the average harvest was 160
million pounds of oyster meat. C. MacKenzie, Jr., "History of Oystering in the United States and Canada Featuring the Eight Greatest Oyster Estuaries," *Marine Fisheries Review* 58, 4 (1996): 1–78.
In 2012 the national harvest was 23,833,089 pounds, or 14 percent of the former peak. www.st.nmfs.noaa.gov/st1/commercial/
landings/annual_landings.html.

12 Unlike the eastern oyster, which has declined by at least 80 percent: The figure comes from The Nature Conservancy's analysis of
global oyster reefs. See M. W. Beck et al., *Shellfish Reefs at Risk: A
Global Analysis of Problems and Solutions* (Arlington, VA: The Nature Conservancy, 2009).

12 Over the course of the last fifty years Americans came to eat more
shrimp: Jack Rudloe and Anne Rudloe, *Shrimp: The Endless Quest
for Pink Gold* (Upper Saddle River, NJ: FT Press, 2010). In this
very thorough account the authors follow the rise in popularity of
shrimp from the 1950s government investment in domestic shrimp
fisheries (p. 39) to shrimp surpassing canned tuna as "America's
best-selling seafood" (p. 41). They also cite a *New York Times* sur-

vey that found that American shrimp consumption per person
nearly quadrupled from 1980 to 2007 (p. 44).

13 **generate more than two hundred million pounds of fish per year:**
The 2012 sockeye salmon run was 29.1 million fish; average size of
sockeye was 5.7 pounds = 166 million pounds of fish in 2012. This
was 28 percent below the 1992–2011 average run of 37.3 million
fish—over 200 million pounds of fish. From the Alaska Department of Fish and Game Division of Commercial Fisheries 2012
Bristol Bay salmon season summary, www.adfg.alaska.gov/static/
applications/dcfnewsrelease/226013052.pdf.

13 **the largest open pit mine in North America placed in dangerous
proximity:** See "The Risks of Pebble Mine" by OurBristolBay at
www.ourbristolbay.com/the-risk-factsheet.html#_ftn2.

13 **79 percent of Alaska salmon is exported:** A summary of Alaska
exports can be found in the report "Alaska Seafood Export Trends"
of the Alaska Seafood Marketing Institute, http://seagrant.uaf
.edu/map/aspli/2011/presentations/exporttrends.pdf.

14 **two-thirds of the salmon that Americans do eat is farmed and
comes to us from abroad:** Pages 123–134 in G. Knapp, C. Roheim, and J. Anderson, chapter 8, "Overview of U.S. Salmon
Consumption," *The Great Salmon Run: Competition Between Wild
and Farmed Salmon* (Washington D.C.: TRAFFIC North America / World Wildlife Fund, 2007).

14 **"restore and maintain the chemical, physical, and biological integrity of the Nation's waters":** Some of the very first words in Section 101(a) of the 1972 Clean Water Act. The full text as amended
through 2002 can be found here: www.epw.senate.gov/water.pdf.

14 **A single oyster is capable of filtering up to fifty gallons of water
per day:** Oyster filtration rates vary from individual to individual
and watershed to watershed and range from twenty to fifty gallons
a day. From Kurlansky's *The Big Oyster,* page 55. Additional information on oyster filtration can be found in "Incorporating Shellfish Bed Restoration into a Nitrogen TMDL Implementation
Plan" by Richard F. Golen, Esq., Professor of Management,
Charlton College of Business, University of Massachusetts–
Dartmouth.

15 **the Deepwater Horizon disaster:** See pages 60–68 in the blow-by-
blow account of the Deepwater Horizon accident: Peter Lehner, *In
Deep Water: The Anatomy of a Disaster, the Fate of the Gulf, and How
to End Our Oil Addiction* (New York, NY: O/R Books, 2010).

15 **BP could be liable for more than $20 billion:** Lehner, *In Deep*

Water, pages 60–68. Estimates of the potential Clean Water Act liability for BP can be found in Tom Fowler, "BP Faces Bout of Spill Liability," *Wall Street Journal,* February 18, 2013, http://online.wsj.com/news/articles/SB10001424127887323764804578312363372704012.

15 **Louisiana marshes where 75 percent of the northern Gulf's seafood is born:** See http://shreveporttimes.gannettonline.com/gns/wetlands/. In addition, Louisiana wetlands comprise 40 percent of the total wetlands in the contiguous United States, and they contribute between 25 and 30 percent of the total volume of U.S. fisheries (excluding Alaska and Hawaii). From the U.S. Department of the Interior, www.doi.gov/pmb/oepc/wetlands2/v2ch8.cfm.

EASTERN OYSTERS

20 **a locally sourced seafood diet:** From "History of the Fishery Industries of New York" (cited above). Matthews writes that there were 475 million pounds of fishery product landed in New York in 1928, 65 million pounds of which were oysters; 79 percent of this total was consumed in the city. Dividing by a then population of nine million people (quite close to the current population of New York), Matthews arrives at a total per capita consumption of 36.78 pounds.

20 **"Canal Street Plan":** Much of my knowledge of the early history of New York oyster culture came to me through Mark Kurlansky's book *The Big Oyster*; see pages 157–158 for the "Canal Street Plan."

20 **more on oysters than on butcher meat:** This according to Margaret Halsey-Gardiner, executive director of New York City's Merchant's House Museum, as quoted in *Newsday,* June 14, 1999, www.newsday.com/the-oyster-connection-natural-pollution-filters-may-help-revive-1.470930.

21 **tidal estuary straddling Brooklyn and Queens:** Jamaica Bay is not technically part of the Hudson River estuary but forms a separate drainage that is more Long Island than New York City. There is debate in the New York marine conservation community as to whether or not a wild oyster was ever present in the bay. It was, however, a major site for the farming of oysters, and there was even a specific "brand" of oyster called a Jamaica.

21 **Hudson-Raritan estuary:** The Raritan River is one of those urban rivers that most people drive over and fail to notice. In fact, the

Raritan is one of New Jersey's largest rivers by volume. It begins in the mountainous central part of the state and ends where Staten Island sits. Collectively the Hudson and Raritan make up an estuary complex that is often referred to as a single ecological entity.

21 **Harbor School students and Baykeeper were there:** The full consortium included Hudson River Foundation, NY/NJ Baykeeper, U.S. Army Corps of Engineers, Port Authority of New York and New Jersey, Urban Assembly New York Harbor School, Harbor Foundation, Governors Island Preservation and Education Corporation, Hudson River Park Trust, U.S. Environmental Protection Agency, New York–New Jersey Harbor and Estuary Program, New York City Department of Environmental Protection, New York City Department of Parks and Recreation, New York State Department of Environmental Conservation Hudson River Estuary Program, NOAA Restoration Center, Bronx River Alliance, Rocking the Boat, and the Bart Chezar–Bay Ridge Flats Oyster Project.

25 **When a healthy oyster population filters and clears the water, sunlight is able to penetrate into the depths, in turn spurring the growth of several species of amphibious grass:** And vice versa. Eelgrass has the effect of facilitating oyster spat settlement. Eelgrass also provides significant water filtration benefits. Both oysters and eelgrass do better in clearer waters and both are key to a healthy, bountiful estuarine system. http://blog.nature.org/science/2013/06/07/eelgrass-seagrass-shellfish-restoration-science-bo-lusk/ and http://seagrant.uconn.edu/publications/magazines/wracklines/fallwinter07/eelgrass.pdf.

26 **They sequester more carbon than any other known ecosystem:** Food energy is calculated from all parts of the marsh—living and decomposed plants, nutrients in the soil, etc. The living plants, which create primary production through photosynthesis, are fed upon by bacteria and insects that in turn are fed upon by spiders . . . and so on up the food chain. The decomposed plants are fed upon by microscopic worms and other nearly invisible creatures that are fed upon by crabs, mussels, etc., that feed other, larger creatures up the food chain. Meanwhile, new nutrients are arriving from the sea via the tide while other transformed nutrients are taken out into the sea. In essence, very little gets wasted—making the salt marsh a very dense and efficient system. See J. R. Packham and A. J. Willis, *Ecology of Dunes, Salt Marsh and Shingle* (London: Chapman & Hall, 1997). There are also two reasons for high carbon sequestra-

tion: (1) salt marsh plants grow quickly, taking up carbon in the form of carbon dioxide for photosynthesis and storing much of it in the body of the plant from roots to leaves; (2) dead plant matter gets buried in the soil of the salt marsh, which is anaerobic, oxygen poor, so decomposition is very slow and carbon remains stored in the buried detritus. www.habitat.noaa.gov/coastalcarbonsequestration .html.

26 **Three-quarters of all the commercial fish species:** This frequently cited number applies to all estuarine environments of which salt marshes are the critical component. See http://oceanservice.noaa .gov/education/tutorial_estuaries/est02_economy.html.

26 **the Great Dismal Swamp:** The Dismal Swamp Company was founded by Washington and other wealthy Virginians in 1763 to drain a huge span of swampland in Virginia and North Carolina as an investment to create more arable land near water that in turn could be sold for profit. Slaves were used to dig ditches to drain the swamp and harvest cedar but the land remained unproductive for crops. After the Revolutionary War Dutch laborers, who were more familiar with draining wetlands, were brought in, but their efforts also failed. In 1793 a canal was completed connecting Chesapeake Bay and Albemarle Sound, but by then Washington had lost interest and sold his share in 1795 (which he later re-gained). The company finally saw a profit from lumber in the 1810s and became the Dismal Swamp Land Company.

27 **wild oyster reefs rank among the most endangered ecosystems:** From Beck et al., *Shellfish Reefs at Risk: A Global Analysis of Problems and Solutions,* The Nature Conservancy (cited above).

27 **lost 70 percent of its historical salt marsh:** The United States lost approximately sixteen thousand acres of coastal salt marsh per year between 2004 and 2009, three times higher than the amount of loss that was documented from 1998 to 2004. But this is modest in comparison to the wetlands loss that occurred between the 1950s and the 1970s, during which time five hundred thousand acres of total wetlands (fresh and salt) were lost on an annual basis. Of that, approximately fifty thousand acres per year was salt marsh. It's curious to note that the U.S. seafood deficit begins to be particularly exacerbated exactly during the period of greatest wetlands loss. Today loss continues at a slower rate, though it is most pronounced in the Mississippi delta; this phenomenon is explained more thoroughly in the shrimp chapter of this book. A summary of historical trends in wetlands loss can be found at www.fws.gov/wetlands/

Documents/History-of-Wetlands-in-the-Conterminous
-United-States.pdf. The most recent update to wetlands trends
can be found at www.fws.gov/wetlands/Documents/Status-and
-Trends-of-Wetlands-in-the-Conterminous-United-States
-2004-to-2009.pdf.

27 **once paved more than four hundred thousand acres:** According to
page 6 of the 1887 "Second Report of the Oyster Investigation," a
document prepared by New York fisheries commissioner Eugene
Blackford, there existed in precolonial times 409,186 acres of suit-
able territory. By 1885 there were only 35,000 acres remaining. *The
Second Report of the Oyster Investigation and of Survey of Oyster Territory: For the years 1885 and 1886* (Albany, NY: Argus Company
Printers, 1887).

28 **female oyster can produce up to one hundred million eggs:** Oyster
promoter extraordinaire Jon Rowley of Seattle likes to say of oysters that they are "her today, guy tomorrow," but in fact it's the
other way around. Oysters are "protandric," meaning they begin
life as males and produce sperm during their first year, then transform into females the following year and stay that way for the rest
of their lives. Oyster fecundity has been long known but historically underestimated. As long ago as 1883 a *New York Times* review
of a presentation on oyster fecundity by a professor Huxley in London reported that the professor had counted one million eggs in a
single mature female, something the *Times* reporter noted would
"break the heart of Malthus." Malthus's heart would have been
further broken by the NOAA's P. S. Galtsoff, who put the number
at one hundred million eggs. See www.nefsc.noaa.gov/publica-
tions/classics/galtsoff1964/c14p306-314.pdf.

Also see Eastern Oyster Biological Review Team, "Status Re-
view of the Eastern Oyster (*Crassostrea virginica*)," Report to the
National Marine Fisheries Service, Northeast Regional Office.
NOAA Tech. Memo. NMFS F/SPO-88, February 16, 2007.

28 **A salmon, by comparison, lays only about three thousand to ten
thousand:** The number of eggs per individual female depends
greatly on size, age, and species. Generally the bigger or older the
salmon, the greater the number of eggs. Chinook salmon average
around ten thousand eggs per fish but sockeye average around
three thousand per fish. Atlantic salmon are in the middle, seven
thousand to eight thousand. Quite a bit less than oysters, in any
case. See C. Skaugstad and B. McCracken, "Fecundity of Chinook
Salmon, Tanana River, Alaska," Fishery Data Series No. 91-8.

Alaska Department of Fish and Game, 1991; and G. B. Pauley, R. Risher, and G. L Thomas, "Species Profiles: Life Histories and Environmental Requirements of Coastal Fishes and Invertebrates (Pacific Northwest)—Sockeye Salmon," U.S. Fish Wildl. Serv. Biol. Rep. 82(11.116), U.S. Army Corps of Engineers, TR EL-82-4, 1989; and C. Fay et al., "Status Review for Anadromous Atlantic Salmon (*Salmo salar*) in the United States," Report to the National Marine Fisheries Service and U.S. Fish and Wildlife Service, 2006.

28 **drop bottomward:** It is believed the settling of oyster larvae on appropriate substrate is cued by a combination of temperature, salinity, a sensitivity to habitat with complex surfaces, and chemicals released by other oysters and biological substances on potential surfaces. See page 7 in Eastern Oyster Biological Review Team, "Status Review of the Eastern Oyster (*Crassostrea virginica*)," Report to the National Marine Fisheries Service, Northeast Regional Office, NOAA Tech. Memo. NMFS F/SPO-88, February 16, 2007.

29 **double the number of fish a given area of water:** A recent NOAA study of oyster reef restoration in Mobile Bay, Alabama, showed increase of blue crab by 297 percent, red drum by 108 percent, spotted sea trout by 88 percent, and flounder by 79 percent after oyster restoration. See www.habitat.noaa.gov/pdf/RAE_fisheries.pdf. From S. B. Scyphers, S. P. Powers, K. L. Heck Jr., and D. Byron, "Oyster Reefs as Natural Breakwaters Mitigate Shoreline Loss and Facilitate Fisheries," *PLoS One* 6, 8 (2011): e22396.

29 **puzzle that stretched from Red Hook in Brooklyn a quarter mile out to sea:** Ellis and Liberty Islands were two of the three islands in New York Harbor known as Oyster Islands, and archeologists have found evidence on both islands of Native American oyster harvesting. www.nps.gov/elis/historyculture/places_oyster_island.htm and http://nynjbaykeeper.org/resources-programs/oyster-restoration-program/oyster-restoration-research-partnership/.

29 **mined out at progressively faster rates:** Oysters were not only harvested for food. The Dutch buried the lime-rich shells in fields where the lime would "sweeten the soil" for crops, used the shells as road pavement or filler, and also burned the shells to create lime paste for home and building construction. These various uses were continued by the English. Pages 12, 56, and 62 in Kurlansky's *The Big Oyster* (cited above).

31 **entrepreneurs began importing juvenile "seed" oysters from still**

functioning wild beds: Seeding in New York waters may have begun as early as 1808 but certainly was becoming common by the 1820s and was mandatory by the 1850s to keep up with growing demand. From M. X. Kirby, "Fishing Down the Coast: Historical Expansion and Collapse of Oyster Fisheries Along Continental Margins," *Proceedings of the National Academy of Sciences* 101, 35 (2004): 13096–13099.

31   its fermentation generating carbonic acid and ammonia waste: Page 86 in Waldman's *Heartbeats in the Muck* (cited above).

32   the presence of oysters in New York restaurants rose continually from the 1860s: From a conversation with Carolyn Hall, a historical marine ecologist and consultant with the Wildlife Conservation Society's New York Seascape Program. See www.wcs.org/saving-wild-places/ocean/new-york-seascape.aspx.

32   1.4 billion oysters a year: A summary of oyster production around the turn of the century can be found at www.forbes.com/sites/nadiaarumugam/2012/02/27/oysters-are-back-in-new-york-city-waters/.

32   consistent sources of oysters until the 1920s: Because Jamaica Bay became a dumping ground for sewage and industrial waste, the famous oysters were beginning to make consumers ill in the early 1900s. All commercial shellfishing in the bay ceased in 1921. Pages 47 and 62 in Frederick R. Black, "Jamaica Bay: A History," Cultural Resource Management Study No. 3 (Washington, D.C.: U.S. Department of the Interior, 1981).

32   a primeval suspicion of the dangers of eating oysters: Some of the oldest dietary laws tell us to avoid bivalves. In Leviticus, God instructs Moses, "Whatsoever hath fins and scales in the waters, in the seas, and in the rivers, them shall ye eat . . . And all that have not fins and scales in the seas, and in the rivers, of all that move in the waters, and of any living thing which is in the waters, they shall be an abomination unto you." The word "abomination" in Hebrew is *toevah* and implies not so much intrinsic evil as ritual uncleanliness. And the word is repeated four times in this passage, twice the number of times in relation to seafood as to landfood. Later commentators on Leviticus made the strictures even more strict. Fish that have scales that are not readily detachable from the skin, like ocean pout and wolffish, or fish that broadly speaking have a snaky/eely appearance, are excluded from most lists of kosher fish. "Maimonides speaks about different qualities of the fish whether they are higher up in the ocean or lower in the ocean,"

Rabbi Menachem Genack, an expert in kosher laws, told me recently. Things that are lower down, bottom-feeders and the like, Genack explained, often get the thumbs-down. This tracks well with the only other category of animals that gets as much abomination associated with them as shellfish: things that creep. "Whatsoever goeth upon the belly," God tells Moses, "creeping things that creep upon the earth, them ye shall not eat; for they are an abomination."

32 **Pacini first discovered the bacillus:** Robert Koch is credited by Kurlansky and others with cholera's discovery in 1884, but Pacini's publication following a European pandemic of the disease in 1854 preceded Koch's publications on the subject by nearly three decades. See www.ph.ucla.edu/epi/snow/firstdiscoveredcholera.html.

33 **hepatitis A and B were found to be present in oysters:** As an aside, my father reports that when he was a medical resident at New York's Bellevue Hospital in the 1950s shellfish poisoning was a common (and much feared) presence in the hospital's intake wards. Symptoms from shellfish poisoning were notoriously random and pernicious and confounded many a young diagnostician.

33 **Pure Food and Drug Act was passed:** Upton Sinclair's great muckraking meat industry exposé *The Jungle* (New York: Doubleday, Jabber, 1906) is widely seen as a catalyst for the passage of the act. One hundred bills had been introduced in Congress to regulate food and drugs. Once established, the newly founded Food and Drug Administration's first primary investigator, Walter G. Campbell, included oysters in a list of products that were to receive "special investigations." See www.fda.gov/AboutFDA/CommissionersPage/ucm113609.htm.

33 **By 1896 New York was shipping a hundred thousand barrels of oysters a year to England:** In the season of 1880–1881 the shipment of oysters to Liverpool, England, from New York was over sixty-eight thousand barrels—and that didn't include barrels shipped to London, Cardiff, or Bristol. Also of note, English oyster eaters preferred small oysters, which meant harvesting New York oysters at one year old or younger. Even at the time, oyster growers were concerned for the future of such early harvest because so many were taken before spawning. Still, economics won, people believed that cultivation would outproduce demand, and the trade of immature oysters continued. From pages 131–134 of the incredibly comprehensive Ernest Ingersoll, *The Oyster-Industry: The History and Present Condition of the Fishery Industries*, a special report

for the Tenth Census of the United States (Washington, D.C., Government Printing Office, 1881).

33 **A Chicago politician ate New York oysters and promptly died:** In 1924 typhoid cases in Chicago, New York, and Washington, D.C. were all connected to New York City oysters—a strong reason for the collapse of the New York oyster industry. T. Andersen, *This Fine Piece of Water: An Environmental History of Long Island Sound* (New Haven: Yale University Press, 2002), 97–98.

34 **one of the hints of replacing a domestic source of seafood:** "Jamaica Bay, Foul with Sewage, Closed to Oyster Beds; 300,000 Bushels Gone," *New York Times*, January 30, 1921.

34 **New York's health commissioner Royal S. Copeland:** Ibid.

35 **production had plummeted to a mere 1 percent:** This figure comes from John M. Kochiss, *Oystering from New York to Boston* (Middletown, CT: Wesleyan University Press, 1974).

35 **mercury-laced pulp effluent from West Coast paper mills:** Oystermen were some of the first citizens to actively pursue polluters before the era of the Clean Water Act. A landmark case occurred in 1968 in Washington State's Olympic Peninsula when oystermen sued Rayonier Inc. for $224,488 in U.S. District Court for damages sustained to their beds in 1941 and 1942. Today, thanks to the closure of the plant and rigorous water quality enforcement by the oyster industry, the Pacific Northwest is one of the United States' biggest oyster-producing regions. See pages 122–123 in E. N. Steels, *The Rise and Decline of the Olympia Oyster*, for The Olympia Oyster Growers Association (Elma, WA: Fulco Publications, 1957).

35 **In Jamaica Bay four wastewater treatment plants were constructed:** Fifty million gallons of sewage were going into Jamaica Bay by 1917 (page 64 in Black's "Jamaica Bay: A History" [cited above]) and the four treatment plants were built from 1903 through 1952. See the New York City Department of Environmental Protection Web site for other facts about the treatment plants: www.nyc.gov/html/dep/html/wastewater/wwsystem-plants.shtml.

37 **Today there are 730 CSOs throughout New York City:** John Waldman reported this number on page 90 in *Heartbeats in the Muck*. For a comparison, the whole state of New York has a total of 937 CSOs. www.dec.ny.gov/chemical/48595.html.

37 **"Cloacina of all the depravities of human nature":** Thomas Jefferson, letter to William Short, September 8, 1823.

38 **restoring five thousand acres of oyster reef:** In 2009 the U.S.

Army Corps of Engineers set forth eleven goals for the ecological rehabilitation of New York Harbor. Key among them was oyster restoration. According to the Comprehensive Restoration Plan, a goal of five hundred acres of oyster reef was set for 2015 and five thousand acres by 2050. See www.hudsonriver.org/download/ORRP_Fall2010Summary.pdf for the program's plans and goals and see www.hudsonriver.org/download/ORRP_Phase1.2013.pdf for progress to date.

40 **the goal of planting a million trees in New York by 2017:** For information on the Million Trees project see www.milliontreesnyc.org/html/home/home.shtml.

40 **the slow process of trying to bring oysters back:** There were of course other early oyster advocates. Bart Chezar, another early oyster restoration advocate, was the first to put oysters in cages on the Bay Ridge Flats. Chezar has moved on from oysters and now focuses on eelgrass and chestnut tree restoration.

42 **U.S. waterway pollution legislation:** My summary of the Clean Water Act's evolution is drawn primarily from N. William Hines, "History of the 1972 Clean Water Act: The Story Behind How the 1972 Act Became the Capstone on a Decade of Extraordinary Environmental Reform," University of Iowa Legal Studies Research Paper No. 12-12, University of Iowa College of Law, May 5, 2012, downloadable at http://ssrn.com/abstract=2045069.

43 **Rivers caught on fire:** The Cuyahoga River, which runs through Cleveland, Ohio, did catch on fire in the summer of 1969 and made the cover of *Time* magazine. Actually, it was the inches-thick sludge of industrial waste floating on the top of the river— and it wasn't the first time. The river, and Cleveland, became the poster children for the nationwide demand to clean up the country's waters (http://clevelandhistorical.org/items/show/63#.UptgOqXoc0Y and http://www.cleveland.com/science/index.ssf/2009/06/cuyahoga_river_fire_40_years_a.html). Fish kills were occurring in numerous places due to anaerobic conditions often brought on by excessive industrial agricultural fertilizer runoff (high nitrogen content), causing algal blooms that consumed oxygen from the water and blocked sunlight. An extreme example is Lake Erie in the 1960s, when it "was declared 'dead' by the press due to the high levels of nutrients accompanied by excessive growth of algae, fish kills, and anaerobic bottom sediments." From the 1996 Food and Agriculture Organization of the United Nations (FAO) document *Control of Water Pollution from Agricul-*

ture, chapter 3: "Fertilizers as Water Pollutants," www.fao.org/
docrep/W2598E/W2598E00.htm.

43 **nationwide testing of 590 water samples:** An excellent summary of
the achievements of the Clean Water Act by Russell McLen-
don can be found at www.mnn.com/earth-matters/wilderness
-resources/blogs/the-clean-water-act-turns-40#.

43 **87 percent of swordfish sampled had mercury levels well above
government-mandated maximums:** An often quoted statistic—
published with the whole story here: William B. Folsom, Dale M.
Crory, and Karyl Brewster-Geisz, "North America—Swordfish
Fishing," *World Swordfish Fishing: An Analysis of Swordfish Fishing
Operations. Past-Present-Future*, vol. 5, prepared by the Office of
Science and Technology, National Marine Fisheries Service,
NOAA, U.S. Department of Commerce, Silver Spring, Mary-
land, 1997. http://archive.org/stream/worldswordfishfi00unit/
worldswordfishfi00unit_djvu.txt.

43 *United States v. Standard Oil:* Standard Oil won the first round of
this case in a Florida district court on the grounds that accidental
gasoline leakage wasn't "refuse" but rather commercially valuable
industrial material. The Supreme Court reversed the decision, ruling that oil in any form has the same negative effect on our waters
and that "its presence in our rivers and harbors is both a menace to
navigation and a pollutant." The case description can be found
here: http://caselaw.lp.findlaw.com/scripts/getcase.pl?court=us&
vol=384&invol=224.

44 **Levels of dioxins in once commercial seafood species:** Dioxin level
numbers from page 163 in John Waldman's *Heartbeats in the Muck*
(cited above). The New York Harbor oxygen levels were regularly
as low as 1.5 milligrams per liter in summer months during the
heavily polluted early twentieth century. After the Clean Water
Act, the average level improved from around 3 milligrams per liter
to 5 milligrams per liter within ten years (http://water.epa.gov/
type/oceb/nep/upload/2007_05_09_oceans_nepccr_pdf_nepccr
_nepccr_ne_parti.pdf). Information on oyster gardening site requirements including oxygen found on the Government of
Virginia site: www.deq.virginia.gov/Portals/0/DEQ/CoastalZone
Management/gardenstep1.pdf.

52 **a fax arrived at Baykeeper's office:** Baykeeper's point of view on
the Keyport raid can be found in the press here: http://atl.gmnews
.com/news/2010-08-19/Front_Page/A_sad_day_for_Raritan
_Bay.html and the New Jersey Department of Environmental Pro-

tection press release on reasons for the removal can be found here: www.state.nj.us/dep/newsrel/2010/10_0053.htm.

57 **The great oyster tamer's name:** My knowledge of the early days of oyster culture came to me through an interview with Ronald Goldberg, chief of the culture/habitat branch at the Milford Shellfish Laboratory in Milford, Connecticut, who was in residence about a decade after Loosanoff left the lab. The stories he related to me were passed on by others who actually worked with Loosanoff. He pointed out to me that Loosanoff is a crucial link in a chain that stretches back to the 1800s. "Things don't happen in a vacuum," Goldberg told me. "Science builds on previous science. In the late 1800s [John] Ryder, [William] Brooks, and Lt. Francis Winslow had a vision that oysters could be farmed. Much field experimentation was undertaken to propagate oysters using primitive equipment and showed that it could be done. Brooks wrote a book called *The Oyster* in 1891 that described oyster embryology. The next step I see was [taken] in the 1920s with Paul Galtsoff. He was born in Russia into a wealthy family and had formal academic training. After the Revolution he emigrated to the United States. He directed an oyster research program for the U.S. Bureau of Fisheries in the early 1920s and then wrote a book called *The American Oyster*. He started writing it early in his career and it was finally published in 1964. The story in Milford starts after Galtsoff moved to Woods Hole and he started assigning scientists to Milford. It was Galtsoff who hired Loosanoff in 1931."

60 **Rhode Island oyster grower Bob Rheault:** Bob "Skid" Rheault was president of Moonstone Oysters in Narragansett, Rhode Island, for twenty-six years. He is currently the executive director of the East Coast Shellfish Growers Association.

60 **home to a thousand baby lobsters, thousands of juvenile sea bass, and other fish:** The scientists also documented that the diversity around the oyster cages was as good as, or better than, nearby federally protected "essential fish habitat" eelgrass beds. Bob Rheault said they found "hundreds (sometimes thousands) more fish and critters than the nearby eelgrass beds supported."

61 **disputes between oyster growers on the Connecticut coast:** Coastal towns in Connecticut began privatizing seafloor for oyster bed ownership in the early to mid 1800s. This fed the debate between public and private use of coastal land. See J. Opton-Himmel and A. W. Whelchel, "Private Shellfish Grounds in Central Connecticut: An Assessment of Law, Policy, Practice and Spatial

Data," The Nature Conservancy, Long Island Sound Program (Publication #2), New Haven, CT, 2010.

62 **Connecticut oyster grower named J. P. Vellotti:** The variety and local color of American oyster growers are too diverse to catalog in the main body of a book. But it is worth noting that J. P. Vellotti in addition to growing edible oysters in relatively close proximity to New York City was at the time of my visit in the process of restoring and saving the *Laurel*—an 1891 Staten Island–built oyster steamer that previously had been the property of Connecticut's most powerful oyster dynasty, the Blooms of the Tallmadge Brothers Oyster Company. Unfortunately, his good fight to preserve the *Laurel* has failed and another bit of our nation's great oyster history has been lost.

63 **seaweed market is concentrated outside the United States:** China, Korea, and Japan are the top producers of seaweed products, a market of $5.6 billion at the time of this 2002 Food and Agriculture of the United Nations (FAO) document: *Prospects for Seaweed Production in Developing Countries*, chapter 1, "The Seaweed Industry—an Overview," www.fao.org/DOCREP/004/Y3550E/Y3550E00.HTM.

63 **Of the $108 million in mussels Americans eat:** The majority of mussels consumed in the U.S. are grown in Canada and New Zealand (www.discovermussels.com/press-releases/canadian-mussel-exports-us-10-cent-surpass-new-zealand-first-time). As a result, NOAA is investing in domestic mussel aquaculture, starting with a project off the coast in Massachusetts: www.nmfs.noaa.gov/stories/2012/08/08_06_12gloucester_mussels.html.

64 **Steve and Sarah Malinowsky:** Information on Steve and Sarah Malinowski's Fishers Island Oyster Farm can be found at www.fishersislandoysters.com.

66 **Museum of Modern Art in 2009 produced a schema:** A description of the MoMA exhibit can be found at www.scapestudio.com/projects/oyster-tecture/.

67 **the sea level around the island of Manhattan is predicted to rise:** By the 2050s, with just sea level rise, the waters around New York City are predicted to rise up to a foot. If rapid ice melt (glaciers and polar ice caps) is included, that rise could be over two feet. www.dec.ny.gov/energy/45202.html.

67 **About 20 percent of New York City is built on landfill:** And construction upon landfill is everywhere in New York City: Battery Park City, east of Pearl Street to South Street, Governors Island,

Rikers Island, Flushing, Brooklyn . . . the list goes on. This figure appeared in a *New York* magazine article sourced from the New York Department of Sanitation. http://nymag.com/guides/everything/trash/trash-statistics-2013-7/.

68 **In 1911 New York City built a flushing system for the Gowanus:** Much has been written about the beleaguered Gowanus Canal by the Environmental Protection Agency, the New York Department of Environmental Protection, and local press. A detailed account can be found at www.epa.gov/region02/superfund/npl/gowanus/pdf/gowanus_prap.pdf; also www.nyc.gov/html/dep/html/dep_projects/cp_gowanus_canal.shtml; also http://cityroom.blogs.nytimes.com/2011/02/23/under-the-gowanus-canal-flushing-out-the-stench/?_r=0.

69 **Between 1830 and 1855 the tonnage of the average vessel entering New York Harbor quadrupled:** Pages 118–120 in Waldman's *Heartbeats in the Muck* (cited above). New York Harbor waterways tend to return to the natural twenty-foot depth when dredging can't keep up with silt from the Hudson. In 1979 the Hudson River Channel hadn't been dredged in five years and was only twenty to thirty feet deep instead of the necessary forty-eight feet. R. F. Shepard, "Who Keeps Our Harbor Channels Deep?" *Seaport* 13, 3 (Fall 1979): 8–15.

70 **"Stevens Institute's hydrodynamic water circulation model":** A water circulation model of the Newark Bay Complex—part of the NY-NJ Harbor—was made between 2002 and 2005 to "determine the relationship between hydrodynamic forcing and contaminant transport pathways." In other words, to see how the water flows from the rivers and through the bay and how various industrial wastes are suspended and carried in the waters and along which paths. See Anne M. Pence, Michael S. Bruno, Alan F. Blumberg, Nadia Dimou, and Kelly L. Rankin, "Hydrodynamics Governing Contaminant Transport in the Newark Bay Complex" (Stevens Institute of Technology, Hoboken, NJ) from Proceedings of the Third International Conference on Remediation of Contaminated Sediments, 2005, www.stevens.edu/ses/ceoe/fileadmin/ceoe/pdf/alan_publications/AFB099.pdf.

73 **Off Governors Island on the Bay Ridge Flats:** See the Oyster Restoration Feasibility Study summary, www.hudsonriver.org/download/ORRP_Fall2010Summary.pdf.

75 **Soundview had been the most successful:** See the Oyster Restoration Research Project (ORRP) Final Technical Report, "ORRP

Phase I: Experimental Oyster Reef Development and Performance Results," www.hudsonriver.org/download/ORRP_Phase1.2013 .pdf.

75 **dermo spread with it:** Dermo can be found from Maine to Florida on the Atlantic coast and all along the Gulf coast. On the East Coast it first appeared in Chesapeake Bay and for the most part remained contained there until the 1990s when it spread both north and south probably in combination with warmer winter temperatures, drought conditions, and accidental contamination from oyster shell waste. For more facts, see the Virginia Institute of Marine Science Dermo fact sheet: www.vims.edu/research/departments/eaah/programs/shellpath/Research/perkinsus_marinus/index.php.

76 **the Olympia oyster:** An excellent history of the Olympia oyster is Rowan Jacobsen's *The Living Shore* (New York: Bloomsbury, 2009). For more information on MSX, see the National Exotic Marine and Estuarine Species Information System (NEMESIS), www.pac.dfo-mpo.gc.ca/science/species-especes/shellfish-coquillages/diseases-maladies/pages/hapneloy-eng.htm. Also see the Virginia Institute of Marine Science MSX fact sheet, www.vims.edu/research/departments/eaah/programs/shellpath/Research/msx/.

83 **Seaport City:** Not only did Bloomberg propose this six months after Hurricane Sandy (www.crainsnewyork.com/article/201306 11/NEWS/130619965); in his last months in office he actively sought proposals to move forward with this extreme plan to shield downtown waterfront property (www.crainsnewyork.com/article/20130731/REAL_ESTATE/130739958#).

SHRIMP

91 **one exception to this trend:** For foundational non-field-based research on the great sweep of the shrimp story, I relied on two principal sources: Jack Rudloe and Anne Rudloe, *Shrimp: The Endless Quest for Pink Gold* (Upper Saddle River, NJ: FT Press, 2009), and Jim Carrier, "All You Can Eat," *Orion*, March/April 2009, www.orionmagazine.org/index.php/articles/article/4395/. The Rudloes run the Gulf Specimen Marine Laboratory in Panacea, Florida, where they conduct research, educate the public, and promote protection of the marine environment. www.gulfspecimen.org.

91 **almost a menu category in and of itself:** In fact, in the last ten years the National Marine Fisheries Service has begun breaking out shrimp as a separate, highlighted category in the executive summary of its annual report on seafood imports and exports.

92 **U.S. per capita consumption:** In 2012, shrimp per capita consumption was 3.8 pounds, slightly less than canned tuna (2.4 pounds) plus salmon (2.0 pounds). From www.aboutseafood.com/about/about-seafood/top-10-consumed-seafoods.

92 **Today 90 percent of our shrimp is farmed and imported:** From NOAA's Office of Science and Technology for January through September of 2013, www.st.nmfs.noaa.gov/apex/f?p=169:2, and by the FAO Fisheries and Aquaculture Department, "The State of World Fisheries and Aquaculture," Food and Agriculture Organization of the United Nations, Rome, Italy, 2012.

96 **When our nation first started to encounter shrimp:** For an overview of the Chinese shrimping industry in San Francisco, see pages 28–31 in the Rudloes' *Shrimp: The Endless Quest for Pink Gold* (cited above). Other facts about the Chinese shrimp camps— particularly the longest existing one, China Camp—can be found in Peter D. Schulz, "Work Camps or Ethnic Villages? The Chinese Shrimp Camps of San Francisco Bay," *Proceedings of the Society for California Archaeology*, 9 (1996): 170–178.

97 **coastal provinces in the Pearl River delta:** For more history, see also Paul Bonnot, "The California Shrimp Industry," *Fish Bulletin No. 38*, State of California, Department of Fish and Game, Scripps Institution of Oceanography Library, 1932. http://content.cdlib .org/view?docId=kt3f59n68z&brand=calisphere&doc.view= entire_text.

97 **A few were sold fresh to white settlers as "San Francisco cocktails." The majority were exported to China:** For statistics on shrimp harvest from the Chinese shrimp camps, see Bonnot's "The California Shrimp Industry" (cited above) and Paul Reilly, Kevin Walters, and David Richardson, "Bay Shrimp," *California's Living Marine Resources: A Status Report*, California Department of Fish and Game, December, 2001.

98 **Washing down with the sediment were toxic materials involved in gold processing:** One of the most-cited assessments of tidal marshes in San Francisco Bay since the gold rush is by Brian F. Atwater et al., "History, Landforms, and Vegetation of the Estuary's Tidal Marshes," in T. J. Conomos, ed., *San Francisco Bay: The Urbanized*

Estuary (San Francisco: California Academy of Sciences, 1979), 347–444. And a current study on floods washing mercury-laden sediments down rivers to the San Francisco Bay from old gold mines is by Michael Bliss Singer et al., "Enduring Legacy of a Toxic Fan via Episodic Redistribution of California Gold Mining Debris," *Proceedings of the National Academy of Sciences*, published ahead of print October 28, 2013, doi:10.1073/pnas.1302295110.

98 **another destructive industry, salt making:** Before colonization, Native Americans harvested salt from the bay, but the commercial enterprise began in 1854 and was in full swing by the late 1860s. A 2002 California Research Bureau presentation on the salt production industry and bittern, the saline liquid byproduct that is especially toxic to grasses and wildlife, is here: www.southbay restoration.org/Cargill%20background%20report.html#_ednref4.

98 **By the mid-twentieth century, the marshes of the bay had been greatly reduced:** See Paul Reilly, Kevin Walters, and David Richardson's "Bay Shrimp" (cited above).

99 **five hundred thousand acres of tidal marshes:** Figures from the Senate Select Committee on Baylands Acquisition, June 20, 2002. See www.southbayrestoration.org/Cargill%20background% 20report.html.

99 **Swamp Land Acts:** Fifteen states were granted "reclamation" rights to swampland, including Louisiana and California. For more information on the acts, see the write-up by the U.S. Geological Survey at www.npwrc.usgs.gov/resource/wetlands/ uswetlan/century.htm.

99 **were all significantly drained during this time:** An excellent summary of the loss in American wetlands can be found in Thomas E. Dahl and Gregory J. Allord, "History of Wetlands in the Conterminous United States," *National Water Summary—Wetland Resources: Technical Aspects*, USGS Supply Paper 2425 (1994): 19–26, www.fws.gov/wetlands/Documents/History-of-Wetlands-in -the-Conterminous-United-States.pdf.

100 **Chinese had laid the essential groundwork:** The most significant contribution the Chinese made to American shrimping was their preservation method of drying platforms. This preceded canning and allowed shrimp to be an exportable commodity. See www .louisianafolklife.org/LT/Articles_Essays/creole_art_shrimping _overv.html.

100 **the largest wild oyster reefs remaining in the United States:** The 2012 landings of Eastern oysters for Louisiana were 5,104 metric

tons; the total for the United States was 10,824. http://www.st
.nmfs.noaa.gov/pls/webpls/FT_HELP.SPECIES.

101 **destination for nineteenth-century Chinese:** A write-up of Brasseaux's research into Chinese shrimpers can be found online at
http://ultoday.com/node/2560.

102 **river prawn that had once migrated:** *Macrobrachium ohione* is a variety of what are called caridean shrimp, which make fresh water
their primary residence. An odd-looking creature with one oversized claw (*macrobrachium* meaning "big arm"), it is nearly transparent in color and, today, just as invisible in its public profile. In
precolonial times, though, it was a major market presence, not just
in the delta region but throughout the river. *M. ohione* made not
just a quick jaunt to the sea from the lower river, as brown shrimp
have, but in fact journeyed more than two thousand miles upstream—all the way north into the upper reaches of the Mississippi's main eastern tributary, the Ohio River. In 1899, two
hundred thousand pounds of *M. ohione* were taken by the poorly
documented fishery of the Mississippi River and we can assume
many more—perhaps orders of magnitude more—were caught
and never reported. Albert "Rusty" Gaudé III in the Louisiana Sea
Grant office confirmed this with me. "I ran the numbers," Gaudé
told me, "and that catch was worth $20 million not in today's dol-
lars but in their money." Multiplied out, that means a catch in today's value of nearly half a billion dollars—more valuable than the
entire U.S. shrimp fishery today.

103 **"a mainstay food in all homes":** This quote is drawn from "Shrimp
Tips from New Orleans," U.S. Fish and Wildlife Service, U.S.
Government Printing Office, 1956.

103 **generate a new shell from a dense layer:** This from Harold McGee,
On Food and Cooking: The Science and Lore of the Kitchen, revised
edition (New York: Scribner, 2004).

104 **"armored heavy-headed little fellow":** "Shrimp Tips from New
Orleans" (cited above).

104 **a man named J. M. Lapeyre of Houma, Louisiana:** The history of
Lapeyre is taken from "The Lapeyre Automatic Shrimp Peeling
Machine, Model 'A,' No. 572, 1979," A National Historic Mechanical Engineering Landmark, American Society of Mechanical Engineers, Biloxi, MS, September 25, 2004.

105 **techniques advanced with the advent of canning:** From William
L. Hobart, ed., "Baird's Legacy: The History and Accomplishments of NOAA's National Marine Fisheries Service, 1871–1996,"

NOAA Technical Memorandum NMFS-F/SPO-18, NMFS Scientific Publications Office, Seattle, WA, June 1996.

106 *Mare Liberum*: A translation of the *Mare Liberum* can be found here: http://oll.libertyfund.org/?option=com_staticxt&staticfile= show.php%3Ftitle=552&Itemid=27#toc_list.

107 **a global standard distance of three nautical miles from shore:** Known as the "cannon shot rule." There were various territorial laws passed concerning the right of the United States to board foreign vessels within twelve nautical miles (1799) or to proprietary rights to oil and gas exploration (1945), but national "ownership" of coastal waters and continental shelf rights weren't officially expanded until 1976. www.nauticalcharts.noaa.gov/staff/law_of _sea.html.

108 **unilaterally declared a two-hundred-mile exclusive:** In 1973 the United Nations began convening meetings around what would eventually be ratified in 1982 as the United Nations Convention on the Law of the Sea, or UNCLOS. The convention established territorial sovereignty to all nations out to two hundred nautical miles from shore. But before UNCLOS could even be fully written down on paper, the United States, always resistant to sign international treaties, unilaterally declared its own two-hundred-mile limit through the Magnuson-Stevens Act. A good summary of the history of Magnuson-Stevens can be found at www.nmfs.noaa .gov/stories/2011/20110411roadendoverfishing.htm. It's notable that to this day the United States refuses to be a signatory to the UNCLOS agreement.

108 **to lower the burden of investment in fisheries:** After the passage of the Magnuson-Stevens Act, the patriotic "pro-growth" attitude promoted expansion of the fishery fleets to such an extent that by the mid-1980s fisheries were often overcapitalized, meaning more money was invested in boats, gear, and fishermen than the populations of fish could sustain—often two or three times more. By the 1990s, restrictions were being put into place to reduce fishing fleets, causing many overinvested fishermen to experience serious economic loss. Michael L. Weber, *From Abundance to Scarcity: A History of U.S. Marine Fisheries Policy* (Washington, D.C.: Island Press, 2002).

109 **"seafood fraud":** A 2012 study by the nonprofit Oceana found that seafood fraud varied from state to state but was consistently in excess of the 25 percent mark. Common species subject to fraud are red snapper, Atlantic cod, and wild salmon. See http://oceana.org/

en/our-work/promote-responsible-fishing/seafood-fraud/learn
-act/national-seafood-fraud-testing-results-map.

109 **Red Lobster, spawned in Lakeland, Florida, in 1968:** When bought by General Mills, Red Lobster had five restaurants. General Mills expanded the chain nationwide, and today the Web site fact sheet claims nearly seven hundred locations in the United States and Canada. www.redlobster.com/press/fact_sheet/.

110 **Fishermen overbuilt to make up for loss in per unit value of seafood:** For a full overview of the rapid rise and subsequent overcapitalized and overfished fall of the fisheries from the passage of the Magnuson-Stevens Act to the government restrictions of the 1990s, including details on various specific fisheries, see Weber's *From Abundance to Scarcity: A History of U.S. Marine Fisheries Policy* (cited above).

111 **turtle excluder device:** Louisiana Sea Grant describes how these devices work online at www.seagrantfish.lsu.edu/management/TEDs&BRDs/teds.htm.

111 **bycatch reduction devices:** Red snapper, a much prized and valued food and sport fish in the Gulf, has been the primary motivation for the use of these devices. See http://masglp.olemiss.edu/Water%20Log/WL18/brds.htm.

113 **"Cancer Alley":** This is about a one-hundred-mile corridor between Baton Rouge and New Orleans that has an incredibly high concentration of industrial facilities, oil refineries, and toxic waste dumps. For a recent story on the area, see Julie Cart, "A Strong Voice in Louisiana's Cancer Alley," *Los Angeles Times*, August 27, 2013.

118 **firmly defined natural hedgerow of mangrove:** A more detailed description of the shrimp boom's social and ecological effects on the developing world can be found in Kennedy Warne, *Let Them Eat Shrimp: The Tragic Disappearance of the Rainforests of the Sea* (Washington, D.C.: Island Press, 2011). In 2010, the International Union for Conservation of Nature (IUCN) warned that two of six mangrove species were in danger of extinction and that over the last sixty years 80 percent of mangroves in India and Southeast Asia have been lost owing to coastal development, climate change, logging, and agriculture; see http://mangroveactionproject.org/issues/mangrove-loss/extinction-threat-for-mangroves-warns-iucn. Not only do mangroves provide a rich ecosystem for coastal marine life, they also, much like oyster reefs, can provide protection to coastal communities from storms. In fact, after the devasta-

tion of 2013 Super Typhoon Yolanda, the Philippines Department of Environment and Natural Resources earmarked about $8 million to restore the coast with mangroves and beaches for just this reason; see www.gov.ph/2013/11/27/denr-sets-aside-p347m-for -coastal-forest-rehabilitation-in-eastern-visayas/.

118 **sweep of shrimp ponds:** See N. J. Stevenson, "Disused Shrimp Ponds: Options for Redevelopment of Mangrove," *Coastal Management* 25, 4 (1997): 423–425, for a discussion about shrimp farming growth, disused shrimp ponds, and the possibility of restoring mangroves in Asia and Latin America.

119 **literacy in Vietnam:** UNICEF reports that literacy for fifteen-to-twenty-four-year-olds from 2007 to 2011 in Vietnam was 96 to 97 percent (www.unicef.org/infobycountry/vietnam_statistics.html). The 1945 literacy rate of 10 percent was reported in the paper "Literacy in Vietnam: An Atlas" by Tram Phan, Ayse Bilgin, Ann Eyland, and Pamela Shaw of the Asia-Pacific Research Institute Macquarie University, http://stat.mq.edu.au/Stats_docs/research_papers/2004/Literacy_in_Vietnam_-_an_atlas.pdf.

120 **Shrimp accidentally wandered into these coastal ponds:** This history of early shrimp farming in Thailand comes from George Chamberlain, "History of Shrimp Farming," chapter 1 in *The Shrimp Book*, ed. Dr. Victoria Alday-Sanz (Nottingham, UK: Nottingham University Press, 2010).

121 **a certain specialty dish emerged:** This fact and a description of the dancing shrimp comes from Carrier's "All You Can Eat" (cited above).

122 **kuruma prawn—an animal that sold live retails today for as much as a hundred dollars a pound:** Value also taken from Carrier's "All You Can Eat" (cited above) and www.shrimpnews.com/Free ReportsFolder/GeneralInformationFolder/FarmedSpecies.html. Dr. Motosaku Fujinaga, the inventor of shrimp cultivation, also wrote about the kuruma being a luxury food in the 1930s and 1940s. So much so that when his lab was destroyed by a typhoon in 1942, because it was the second year of World War II and Japan was suffering from food shortages, the country did not support the immediate rebuilding of his lab or further research on the shrimp for many years. From Motosaku Fujinaga (Hudinaga), "Kuruma Shrimp (*Litopenaeus japonicus*) Cultivation in Japan," *Proceedings of the World Scientific Conference on the Biology and Culture of Shrimps and Prawns*, Food and Agriculture Organization, 1968 (www.fao.org/docrep/005/ac741t/ac741t15.htm).

122 Japanese marine biologist named Motosaku Fujinaga: Carrier's "All You Can Eat" (cited above) and *The Shrimp Book* (also cited above). Of particular use was George Chamberlain's chapter, "History of Shrimp Farming."

124 He had a very simple dream, the graduate student recalled: "To make shrimp an affordable food": From Chamberlain's chapter in *The Shrimp Book* (cited above) and from his opening essay "Lessons from the Pioneers," *Global Aquaculture Alliance*, 2005, http://pdf.gaalliance.org/pdf/GAA-Chamberlain-Jun05.pdf. Quotations are attributed to Dr. I Chiu Liao, a 1968 post-doc of Fujinaga, who became the director general of the Taiwan Fisheries Research Institute.

124 It took Fujinaga years to develop the right sequence of food: Marifarms Inc., a U.S. company, purchased the rights to Fujinaga's method and received a patent (No. US3477406 A) in 1969 for the "Method of cultivation of penaeid shrimp" invented by Fujinaga. See www.google.as/patents/US3477406 for a full description of the method.

125 But they soon realized that labor costs would be much cheaper in South America and Asia: In fact, Ralston Purina produced the shrimp feed "Experimental Marine Ration 25" in a company farm in Panama. From Chamberlain's chapter in *The Shrimp Book* (cited above). Also covered in pages 207–210 of the Rudloes' *Shrimp: The Endless Quest for Pink Gold* (cited above).

126 soon after tiger prawn harvests soared: For an early paper on this success, see Alfred C. Santiago Jr., "Successful Spawning of Cultured *Litopenaeus monodon* Fabricus After Eyestalk Ablation," *Aquaculture* 11, 3 (July 1977): 185–196.

126 they don't have much in the way of an immune system and are notably susceptible to disease: See P. Roch, "Defense Mechanisms and Disease Prevention in Farmed Marine Invertebrates," *Aquaculture* 172 (1999): 125–145.

126 powerful antibiotics nitrofuran and chloramphenicol: Unfortunately, chloramphenicol is one of the most effective antibiotics against vibriosis, so it is still in use. See S. Kitiyodom, S. Khemtong, J. Wongtavatchai, and R. Chuanchuen, "Characterization of Antibiotic Resistance in *Vibrio* spp. Isolated from Farmed Marine Shrimps (*Litopenaeus monodon*)," *FEMS Microbiology Ecology*, 72: 219–227, 2010. And currently the U.S. Food and Drug administration has an "import alert" for seafood containing residues of nitrofuran from Bangladesh, India, Indonesia, and Malaysia. www.accessdata.fda.gov/cms_ia/importalert_31.html.

127 Even in that small sampling, violations occur: Only 2 percent of all imported seafood shipments are inspected, but those from manufacturers or countries that are determined to be a higher risk or even flagged with "import alerts" because of previous violations are generally chosen for inspection. Sven M. Anders and Sabrina Westra, "A Review of FDA Imports Refusals—U.S. Seafood Trade 2000–2010," selected paper prepared for presentation at the Agricultural & Applied Economics Association's 2011 AAEA & NAREA Joint Annual Meeting, Pittsburgh, Pennsylvania, July 24–26, 2011.

127 Rediscovered in the late 1980s, it was soon repurposed toward aquaculture: Pages 211–213 in the Rudloes' *Shrimp: The Endless Quest for Pink Gold* (cited above).

128 By 1992 the whiteleg shrimp had spread around the world: *L. vannamei* is farmed in both marine and fresh water and 80 percent is an average for Asian and Latin American farming of the species. In China alone, the proportion is about 85 percent. Diego Valderamma and James L. Anderson, "Shrimp Production Review," Global Outlook for Aquaculture Leadership (GOAL), Santiago, Chile, November 6–9, 2011.

128 hit by wafts of putrid air, causing them all to be engulfed in a wave of nausea: Pages 214–215 in *Shrimp: The Endless Quest for Pink Gold*.

129 good land for a historically much more important subsistence crop: rice: Pages 218–219 in *Shrimp: The Endless Quest for Pink Gold*. Also discussed in the PhD dissertation of Le Canh Dung, "Environmental and Socio-economic Impacts of Rice-Shrimp Farming: Companion Modeling Case Study in Bac Lieu Province, Mekong Delta, Vietnam," Chulalongkorn University, 2008, Bangkok, Thailand.

129 From 1969 to 1990 the country's million acres of mangroves shrank by 33 percent: The earlier statistics are from K. Kathiresan, "Threats to Mangroves: Degradation and Destruction of Mangroves," in *United Nations University Syllabus for "Training Course on Mangroves and Biodiversity,"* 479–481, 2010 (http://ocw.unu .edu/international-network-on-water-environment-and-health/ unu-inweh-course-1-mangroves/Degradation-and-destruction- of-mangroves.pdf). The 2008 mangrove area statistic is from Richard McNally, Angus McEwin, and Tim Holland, "The Potential for Mangrove Carbon Projects in Vietnam," SNV-Nether-

lands Development Organisation REDD+ Programme, Ha Noi, Vietnam, March 2011.

130 **Farm 184:** My understanding of the workings of Farm 184 came from an in-person interview with Le Hoang Vu, director of Ca Mau Province People of Committee 184 Forestry Company.

133 thirty times as much protein as is turned out by an acre of cow grazing land: Intensive grazing can produce as much as 120 pounds of beef per acre. If raw shrimp has 20 grams of protein per pound, then 18,000 pounds have 360,000 grams of protein. And if a raw steak has approximately 100 grams of protein per pound, then 120 pounds has 12,000 grams of protein—30 times less. Shrimp per acre from "All You Can Eat" (cited above). Beef per acre from Melvin R. George, Ronald S. Knight, Peter B. Sands, and Montague W. Demment, "Intensive Grazing Increases Beef Production," *California Agriculture*, 43, 5 (September-October 1989): 16–18.

134 **over 70 percent of all Vietnamese shrimp goes abroad, 90 million pounds of it to the United States alone:** At the NOAA Fisheries Office of Science and Technology Shrimp Import site, totals for shrimp imported by the United States by month or for a calendar year are listed by country: www.st.nmfs.noaa.gov/apex/f?p=169:2: 3341819280453431::NO::.

136 **tripolyphosphates—a common chemical that increases the weight of seafood before it gets to the consumer:** This process and the pros and cons are discussed in Alex Augusto Gonçalves and José Luis Duarte Ribiero, "Do Phosphates Improve the Seafood Quality? Reality and Legislation," *Pan-American Journal of Aquatic Sciences* 3, 3 (2008): 237–247. And here's a paper that studies overuse of enhancers including tripolyphosphate on shrimp of Asian origin: Sujay Paul et al., "Effect of Sodium tri polyphosphate (STPP) and Foreign Materials on the Quality of Giant Freshwater Prawn (*Macrobrachium rosenbergii*) Under Ice Storage Condition," *Food and Nutrition Sciences* 3 (2012): 34–39.

140 **shrimp had been turned into an international commodity:** The analysis can be found in James Anderson and Quentin S. W. Fong, "Aquaculture and International Trade," *Aquaculture Economics and Management* 3 (1997).

140 **directly competes for American resources:** Statistics on U.S. exports of fish meal can be found in the report "Little Fish in Big Demand," Traffic North America, www.traffic.org/species-reports /traffic_species_fish11.pdf.

141 **Asian producers augmented their product lines:** In China, this process of aquaculture diversification actually began during the Tang dynasty (618–907 AD), when the farming of common carp was banned because the Chinese word for common carp sounded like the emperor's family name. Anything that sounded like the emperor's name could not be kept or killed. This led to the polyculture of a much broader range of species. From Herminia R. Rabanal, "History of Aquaculture," ASEAN/UNDP/FAO Regional Small-Scale Coastal Fisheries Development Project, Manila, Philippines, 1988. http://www.fao.org/docrep/field/009/ag 158e/ag158e02.htm.

141 **Vietnamese Pangasius catfish:** More on the Pangasius catfish in my *New York Times Magazine* article "A Catfish by Any Other Name," www.nytimes.com/2008/10/12/magazine/12catfish-t.html?page wanted=all.

141 **China began growing traditional American-born species of catfish on Chinese farms:** The battle to regulate imported catfish continues. An additional level of inspection at the Agriculture Department that was brought into law in 2008 because of Vietnamese and Chinese catfish imports is being challenged. Opponents say there are enough inspections through the FDA and NOAA, but catfish farmers and legislators of the U.S. South argue to the contrary. See Ron Nixon, "Number of Catfish Inspectors Drives a Debate on Spending," *New York Times*, July 26, 2013. China has declared the additional inspection "protectionism" and warns that tariffs on non-seafood U.S. products exported to China could be considered. In "China Calls Catfish Program 'Protectionism,'" *Seafood Source*, August 29, 2013.

142 **In 1980 Professor Fusui Zhang began discussions:** Two articles were sources for this history about Professor Zhang and scallops: Kenneth K. Chew, "Bay Scallop Culture in China Revisited," *Aquaculture Magazine*, May/June 1993; and Mark Allan Lovewell, "New England Bay Scallop Fishery Flourishes Across the Pacific Ocean," *Vineyard Gazette*, 2007.

144 **excruciatingly detailed testimony:** Since the time of this exchange BP won an appeal to cap compensation costs and to more rigorously investigate dubious damage claims. The recording of this testimony can be found at http://www.c-spanvideo.org/program/294728-1.

145 **much more urgent alarm system blaring:** The disappearance of the Louisiana marsh was first brought to my attention just after the BP Deepwater Horizon blowout when Rowan Jacobsen produced

what is perhaps the best overview of the ecological background for the spill in *Shadows on the Gulf* (New York: Bloomsbury, 2011). Jacobsen's work in turn led me to Mike Tidwell's influential and elegiac *Bayou Farewell* (New York: Vintage Departures, 2003). Tidwell's incredibly prescient book was written before both Hurricanes Katrina and Rita as well as before the spill but it manages to foreshadow so much of what has come to pass as a result of all three catastrophes.

145   **The state of Louisiana is currently losing marshland at a rate of a football field every hour:** This is the rate of loss if it occurred constantly according to Brady R. Couvillion et al., "Land Area Change in Coastal Louisiana from 1932 to 2010: U.S. Geological Survey Scientific Investigations Map 3164, scale 1:265,000, 12 p. pamphlet," USGS, 2011.

145   **petroleum exploitation:** To get a visual sense of the vastness of the petrochemical infrastructure laid above and below the Louisiana marsh, readers would be well advised to page through Richard Misrach and Kate Orff's *Petrochemical America* (New York: Aperture, 2012).

146   **James Buchanan Eads was one of the first:** Eads initially opposed the idea of cutting off river bends but was won over to it during a long fight for dominance with his archrival, Andrew Atkinson Humphreys. The details of their conflict are superbly rendered in John M. Barry, *Rising Tide: The Great Mississippi Flood of 1927 and How It Changed America* (New York: Simon & Schuster, 1997).

148   **Today all that fertilizer goes directly into the Gulf:** Algae in the nearshore Gulf of Mexico waters feed on the excess fertilizer nutrients nitrogen and phosphorus that wash down the length of the Mississippi during spring rains. As the waters warm, algae blooms begin and continue expanding until they use up all (or nearly all) the oxygen in the waters. No oxygen, and no sunlight penetrating the algae filled waters, means very little life under the surface—hence the dead zone. For a summary see: http://serc.carleton.edu/microbelife/topics/deadzone/index.html.

152   **BP is potentially liable for a fine of $1,100 to $4,300 per barrel of oil spilled:** The penalty amount per barrel is based on whether BP will be judged guilty of "negligence" or "gross negligence." As of fall 2013 there was a discrepancy of over a million barrels between BP's and the U.S. government's estimate of how many barrels were spilled. That, too, will have huge consequences in determining BP's final tally of fines.

153 **the rebuilding of the Louisiana marsh:** Some $17.9 billion of the $50 billion plan is dedicated to marsh creation—36 percent of the funds. See the details of the plan here: www.lacpra.org/assets/docs/2012MP/Draft_2012_Master_Plan_Low_Res.pdf.

154 **still short of the estimated $60 billion:** This estimate varies but comes to me in this instance from Rowan Jacobsen's *Shadows on the Gulf* (cited above), pages 182–183.

161 **early mortality syndrome:** EMS has also recently spread to Mexico. For an account of the rising shrimp prices see http://triblive.com/state/pennsylvania/4862638-74/shrimp-disease-ems#axz z2nl7w9kUI. And for more information about the disease and losses to the market, see www.gaalliance.org/newsroom/news.php ?Cause-Of-EMS-Shrimp-Disease-Identified-107.

SOCKEYE SALMON

164 **this last great place that produces 5.3 billion pounds of seafood:** This data is from the Alaska Resource Development Council. Salmon is a large part of that product, but the overall dominant seafood from Alaska is walleye pollock, the stuff of fast-food fish sandwiches, and the fake crab in California rolls. See www.akrdc.org/issues/fisheries/overview.html.

164 **a full half billion pounds *more*:** Total U.S. seafood consumption currently ranges from fourteen and a half to sixteen pounds per individual. The amount rose consistently for the last two decades and seems to have plateaued or even declined slightly in the last few years. The 5.3-billion-pound number is from 2011, the latest year for which there is data. *SeafoodSource.com*, September, 18, 2012, www.seafoodsource.com/newsarticledetail.aspx?id=17803.

164 **an eight-hundred-mile pipeline:** A look at the $50 billion gas pipeline project is in Bloomberg News at www.bloomberg.com/news/2012-09-23/alaska-sees-asia-driving-annual-20-billion-via -pipeline.html. A map of the proposed pipeline can be found online at www.arcticgas.gov/alaska-lng-project.

164 **a two-thousand-mile-pipeline:** Presidents Ford and Carter approved the plan and Canada had also provisionally assented, but by the 1980s the pipeline proposal was effectively dead. Today, in light of the potential for domestic gas going to Asia, the powers that be are in the early stages of reconsidering a pipeline to more easily direct shipments to the lower forty-eight. Excellent articles

by Bill White about Alaska's pipeline pursuit can be found online at www.arcticgas.gov/Alaska-gas-pipeline-wars-1971-1982 and www.arcticgas.gov/guide-alaska-natural-gas-projects.

165 **drill into its aquifers:** Alaskan water shipments to the lower forty-eight are prevented through the Jones Act, which prevents the expropriation of resources across state lines. But the Jones Act does nothing to stop the *international* transport of water. Alaskan water export plans are outlined in Maude Barlow and Tony Clarke, *Blue Gold: The Fight to Stop the Corporate Theft of the World's Water* (New York: New Press, 2002).

165 **appears to support serious industrial exploitation of the bay:** Former governor Sarah Palin has seemed to endorse the development of Pebble Mine, though she has been careful in her wording. When an Alaska ballot initiative was introduced in 2008 that would have prohibited new mines from discharging pollutants harmful to humans or salmon, Palin, against state constitution rules, expressed her "personal" opposition to the measure. The ballot initiative was later defeated.

166 **a run that generates as much as two hundred million pounds:** According to the Alaska Department of Fish and Game, since 1990, Bristol Bay harvest has been as much as 243.5 million pounds in one year. In the past five years, harvest has been between 117 million (2012) and 183 million (2009). For harvest figures from 1980 to 2011, see www.adfg.alaska.gov/index.cfm?adfg=commercialbya reabristolbay.main.

172 **less than 3 percent of the rock in the area contains anything of value:** Northern Dynasty breaks down the mineral and tailings percentages on their site. Of the earth removed, 20 percent doesn't contain ore. Of the remaining 80 percent, 3 percent *of that* has minerals (gold, copper, etc.). The other 97 percent of the 80 percent is tailings that will have to be stored somewhere. So, really, only about two and a half percent of the ground that would be mined is a valuable mineral (http://www.northerndynastyminerals.com/i/pdf/NDM_Backgrounder_Sep05.pdf).

173 **loosely designated the state-owned portion:** A full history of the land struggles surrounding Bristol Bay is detailed in a comprehensive account by Geoffrey Y. Parker, "Pebble Mine, EPA, and the History of Federal and State Efforts to Conserve the Kvichak and Nushagak Drainages of Alaska," 2012.

173 **Cominco didn't want to take the strike further:** See http://alaskan sforbristolbay.com/backgrounder.php.

174 **assets in excess of $66 billion:** Anglo American's corporate profile can be found here: http://markets.ft.com/research//Markets/Tearsheets/Financials?s=AAL:LSE&subview=BalanceSheet&period=a.

174 **valued at $500 billion:** For the area of the potential mine, see Stuart Levit and David Chambers, "Comparison of the Pebble Mine with Other Alaska Large Hard Rock Mines," Center for Science in Public Partnership, February 2012.

176 **"You couldn't pick a better place to ruin both drainages":** The Kvichak and Nushagak are the two principal rivers of the Bristol Bay watershed. Between them they account for more than 50 percent of the salmon run in Bristol Bay. Pebble Mine has the potential to affect both vital watersheds.

177 **oldest salmon-like creature:** An excellent source on salmon history is Jim Lichatowich, *Salmon Without Rivers* (New York: Island Press, 2001).

178 **bringing carbon in the atmosphere:** Scientists looking at sodium carbonate concentrations arrived at the 1,125 figure in 2006 (http://news.nationalgeographic.com/news/2006/09/060928-hotearth.html), and in 2009 the high was estimated around 2,240 ppm. Bridget L. Thrasher and Lisa C. Sloan, "Carbon Dioxide and the Early Eocene Climate of Western North America," *Geology* 37 (September 2009): 807–810. But some scientific models put the Eocene number much higher—at around 4,000 ppm. Because the modeling is based on complex ancient data, multiple drivers, and high levels of uncertainty, these numbers are used with caution and many caveats. See Dana L. Royer, "CO_2-forced Climate Thresholds During the Phanerozoic," *Geochimica et Cosmochimica Acta* 70 (2006): 5665–5675. The point is that the Eocene atmosphere had very, very high levels of carbon and Earth was a warmer planet.

178 **possible reducer of inflammation:** Studies have shown the potential benefits of astaxanthin supplements, but experts warn that the studies have been small and short term (www.berkeleywellness.com/supplements/other-supplements/article/astaxanthin-hype). That said, eating wild salmon with astaxanthin is an easy way to benefit from levels existing in nature—no need to turn to supplements.

178 **A single three-ounce portion of cooked wild sockeye salmon contains eight hundred milligrams of omega-3s:** Based on the

USDA National Nutrient Database, http://seafoodhealthfacts
.org/seafood_choices/salmon.php.

179 **the species's difficult-to-master life cycle:** Early attempts at domesticating sockeye salmon were marked by outbreaks of infec-
tious hematopoietic necrosis (IMH), a disease that exists in the
wild but is amplified in the hatchery. Infection results in high
mortality of juvenile fish through destruction of major organs, including the kidneys. In the 1970s, Alaskan egg and fry hatcheries
experienced a 40 percent loss of the incubated sockeye. This led to
rounds of research and eventually new procedures implemented in
1981 that decreased loss to 4 percent. Tim R. McDaniel et al.,
"Alaska Sockeye Salmon Cultural Manual," Special Publication
Number 6, Alaska Department of Fish and Game, August 1994.
As for aquaculture: until very recently there had been no such
thing on the market as a farmed sockeye salmon. Willowfield Enterprises, a food importer, wholesaler, distributor, and manufac-
turer, has developed a landlocked system—"land based" and
"closed containment," no contact with rivers or sea—that in 2013
began marketing farmed Pacific sockeye as West Creek sockeye. A
National Public Radio story first brought this to my attention:
Alastair Bland, "Can Salmon Farming Be Sustainable? Maybe If
You Head Inland," npr.org, May 2, 2013, www.npr.org/blogs/the
salt/2013/05/02/180596020/can-salmon-farming-be-sustainable
-maybe-if-you-head-inland. Willowfield Enterprises Web site:
www.willowfield.net/products.html.

179 **The sockeye's lake fixation:** Sockeye salmon are unique among the
Oncorhynchus genus because some varieties do spend half their lives
in lakes before migrating out to sea. This type is referred to as the
lake type. There is also the sea type that, like other salmon, spends
just weeks or months in freshwater before heading to sea; and the
"kokanee" type that is only a lake dweller—it has no marine part of
its lifecycle. Chris C. Wood, "Sockeye Salmon Ecotypes: Origin,
Vulnerability to Human Impacts, and Conservation Value," in C.
A. Woody, ed., *Sockeye Salmon: Evolution, Ecology, and Management*
(Bethesda, MD: American Fisheries Society, 2007), pages 1–4.

180 **Iliamna, a lake that at any given time may contain more than a
billion sockeye:** The 2012 sockeye salmon run count for Bristol
Bay was 29.1 million adult fish and was below the average of 37.3
million over the last twenty years. Lake Iliamna is the largest lake
drained by the Kvichak River, which historically had runs of 10

million or more, but the prediction for 2013 is 5 million (Douglass M. Eggers, Cathy Tide, and Amy M. Carroll, "Run Forecasts and Harvest Projections for 2013 Alaska Salmon Fisheries and Review of the 2012 Season," Special Publication 13-03, Alaska Department of Fish and Game, February 2013). Lake Clark feeds Lake Iliamna and in the 1980s runs averaged over a million fish. In the early 2000s runs averaged just under 300,000 (Daniel B. Young, "Distribution and Characteristics of Sockeye Spawning Habitats in the Lake Clark Watershed, Alaska," Technical Report NPS/NRWRD/NRTR-2005/338, National Park Service, U.S. Department of the Interior, August 2005). But that's just adult fish. Christopher Boatright, a research scientist with the University of Washington's Department of Aquatic and Fishery Sciences who conducts fieldwork on the Bristol Bay sockeye runs, wrote me, "I looked fairly closely at this claim [of more than 1 billion fish in Lake Iliamna]. I am very comfortable saying it is true. I used data from 1965–2006 (brood year return for Kvichak drainage only complete through 2006), and it is more than reasonable to say that it would be the very rare instance when there would be less than 1 billion sockeye fry (stage between alevin and smolt) in the lake. It appears much more likely that Iliamna and Clark basins in any one year would have greater than 2 billion sockeye fry as opposed to less than 1 billion."

180 **about three hundred million tons of minerals heading out to barges:** Levit and Chambers estimate that Pebble Mine will yield 10.8 billion tons of mineral-holding rock in their "Comparison of the Pebble Mine with Other Alaska Large Hard Rock Mines" (cited above). Three percent of that, or the amount of the ore that is actually copper, gold, or other valuable minerals, is 323 million tons.

181 **As Carl Safina writes:** Carl Safina, *The View from Lazy Point* (New York: Henry Holt, 2011), page 168.

181 **banned large-scale salmon traps:** As part of the Alaska Constitutional Convention held during the transition to statehood, a request to outlaw the enormous fish traps that essentially enclosed the entire mouth of a river to catch salmon was proposed by Eldor Lee, a local fisherman. Most of the giant fish traps were owned by outside interests, not Alaskans, and they monopolized the harvest of salmon. The ban passed and was written into the state constitution. http://www.alaska.edu/creatingalaska/constitutional-convention/delegates/lee/.

181   **a tenet of the state constitution:** See Charles P. Meacham and John H. Clark, "Pacific Salmon Management—the View from Alaska," *Alaska Fishery Research Bulletin* 1, 1 (1994): 76–80.

182   **generating $300 million in local profit and $1.5 billion when all the affiliated business it creates is taken into account:** About one fifth of the economic impact of the Bristol Bay industry stays local. This is because many fishermen, processors, and others in the industry are from other states—including a third of them from Washington state. Gunnar Knapp, Mouhcine Guettabi, and Scott Goldsmith, "The Economic Importance of the Bristol Bay Salmon Industry," Institute of Social and Economic Research, University of Alaska–Anchorage, April 2013.

183   **Bingham Canyon mine in Utah:** The state of Utah filed a Natural Resource Damage Claim, which was settled in 1995. The surrounding contaminated area, including two creeks and the groundwater/drinking water source of the Jordan River, have since been declared Superfund sites. It should be noted that Bingham Canyon is half the size of the proposed Pebble Mine. See Pebble Mine's report on the Bingham Canyon pollution here: http://pebblescience .org/Pebble-Mine/water-impact.html. For links to documents relating to Superfund activities, go to Utah's Department of Environmental Quality Web site: www.deq.utah.gov/businesses/kenne cott/cercla.htm.

183   **will require perpetual on-site remediation:** The exact terminology from "The External Peer Review of EPA's Draft Document, An Assessment of Potential Mining Impacts on Salmon, Ecosystems of Bristol Bay, Alaska" states: "Based on the hypothetical mine scenario, perpetual management of the geotechnical integrity of the waste rock and tailings storage facilities, as well as perpetual water treatment and monitoring, most likely will be necessary." The full report can be found on Northern Dynasty's Web site: http://www.northerndynastyminerals.com/i/pdf/ndm/ FINAL%20Peer%20Review%20Report%20Bristol%20Bay.pdf.

184   **find wild salmon too, well, salmony:** In a 2013 taste test conducted by the *Washington Post*, farmed salmon beat two species of wild salmon (coho and king) hands down. The most preferred was frozen Norwegian salmon that had been packed in a brine solution. A write-up of that taste test can be found at http://articles .washingtonpost.com/2013-09-24/lifestyle/42349449_1_wild -salmon-farmed-seafood-watch.

184   **Two-thirds of the salmon Americans consume is farmed:** From Gunnar Knapp, Cathy Roheim, and James L. Anderson, "The

Great Salmon Run: Competition Between Wild and Farmed Salmon," Traffic North America, World Wildlife Fund, 2007.

186 **79 percent of all Alaska salmon is exported:** A summary of Alaska exports can be found at http://seagrant.uaf.edu/map/aspli/2011/presentations/exporttrends.pdf.

186 **"If you eat the food in China, that will kill you. But if you do not eat the food in China, that will kill you even faster":** This expression was brought to my attention by Jason Czarnezki, the former faculty director of Vermont Law School's U.S.-China Partnership for Environmental Law. The original Chinese saying is 在中国，吃东西会死，不吃东西呢，死得更快。 (transliterated as Zai Zhong Guo, Chi Dong Xi Hui Si, Bu Chi Dong Xi Ne, Si De Geng Kuai).

187 **tainted with the carcinogenic additive melamine:** From Jason Czarnezki, "Are Food Imports from China Safe?," *VT Digger*, November 11, 2011, http://vtdigger.org/2011/11/11/czarnezki-field-are-food-imports-from-china-safe/. A second reference is "Melamine Contamination of Aquaculture to Be Probed," *The China Post*, September 26, 2008, www.chinapost.com.tw/taiwan/national/national%20news/2008/09/26/176228/Melamine-contamination.htm.

187 **bribes that are acknowledged to have been linked to much of the early failure in inspection:** For more on the story about Xiaoyu's execution, see Joseph Khan, "China Quick to Execute Drug Official," *New York Times*, July 11, 2007.

187 **processed imported seafood—fish sticks, crab cakes:** See the USDA guidelines concerning country of origin labeling: www.fsis.usda.gov/wps/portal/fsis/topics/food-safety-education/get-answers/food-safety-fact-sheets/food-labeling/country-of-origin-labeling-for-meat-and-chicken/country-of-origin-labeling-for-meat-and-chicken.

187 **Many in China remain wary of their own food:** As an example, a 2013 story about Chinese demand for powdered milk from European producers can be found online here: www.asianews.it/news-en/China-after-melamine-milk-scandal-powder-milk-smuggled-from-Europe-and-Hong-Kong-27792.html.

188 **maintained the so-called *tegong* system:** Jason J. Czarnezki, Cameron Field, and Yanmei Lin, "Global Environmental Law: Food Safety and China," *Georgetown Environmental Law Review* 26 (2013). The authors write: "In contemporary China, it is the degradation of the environment and a limited supply of healthful food

that is fueling the parallel 'Tegong' food system for the elites. A *Southern Weekend* reporter was able to sneak inside the 'Beijing Customs Administration Vegetable Base and Country Club,' which is a farm to provide organic food to Beijing officials, and published a story on the 'Tegong' system. He Bin, a law professor from China University of Political Science and Law, posted a blog on Caijing sharing his experiences in visiting the 'Tegong (special) farms' established by central ministries and local government agencies when he gave lectures in these places. Grains, vegetables, and meats grown in these special farms are all organic and directly supplied to their contracted agencies. Local protectionism, corruption, lack of resources to detect violations among the fragmented industry, unethical practices, and inadequate tools and mechanisms in responding to stop violations are all root causes for the weak enforcement of environmental laws and food safety laws that have been identified by scholars."

188   **Two-thirds of all Alaska seafood:** A detailed report on the nature of the Alaska seafood business is "Economic Value of the Alaska Seafood Industry," McDowell Group for the Alaska Seafood Marketing Institute, July 2013.

189   **I don't know how to prepare it:** A recent survey by the Institute of Food Technologies found that only a third of consumers were comfortable preparing seafood. See www.ift.org/food-technology/past-issues/2011/april/features/food-trends.aspx?page=viewall.

190   **odorless and virtually tasteless, with very little omega-3 content:** See Elisabeth Rosenthal, "Another Side of Tilapia, the Perfect Factory Fish," *New York Times*, May 2, 2011. A cooked one-ounce portion of tilapia has only 17 percent of the omega-3s that sockeye salmon does (67 mg vs. 399 mg), yet tilapia has nearly three times the omega-6s (84 mg) that sockeye has (31 mg). Nutritional content taken from SelfNutritionData. For sockeye: http://nutritiondata.self.com/facts/finfish-and-shellfish-products/4112/2; for tilapia: http://nutritiondata.self.com/facts/finfish-and-shellfish-products/9244/2.

192   **Until Vice President Biden brokered:** After the health scares and import bans of Chinese products in 2007 and 2008, the FDA increased inspections, but China has been understaffed, foreign inspectors' visas have been denied, and up until 2013 only a third of the inspections were done by FDA non-Chinese inspectors. In December 2013, Vice President Biden confirmed a commitment to

add U.S. inspectors, which will increase the number of inspections eightfold: 20 to 25 per year to 160. See Anna Edney and Drew Armstrong, "China Drugmakers Face U.S. Scrutiny on Investigator Bump," *Bloomberg News*, December 10, 2013.

192 **Air quality on an average day in Beijing:** Though U.S. officials backed away from making this comparison in a recent article in *The Atlantic*, some experts still stated a preference for breathing 9/11 air over Beijing air. See www.theatlantic.com/china/archive/2013/09/is-the-air-quality-in-beijing-worse-than-ground-zeros-after-9-11/279589/.

192 **"so contaminated in 2011 that [their waters] were unsuitable for human consumption":** From Damien Ma and William Adams, "If You Think China's Air Is Bad . . . ," *New York Times*, November 7, 2013, www.nytimes.com/2013/11/08/opinion/if-you-think-chinas-air-is-bad.html?_r=0&hp=&adxnnl=1&rref=opinion&adxnnlx=1383904423-C6e95KTFs5P/Pd38Rerogw. The book from which this op-ed was adapted is Damien Ma and William Adams, *In Line Behind a Billion People: How Scarcity Will Define China's Ascent in the Next Decade* (Upper Saddle River, NJ: FT Press, 2013).

195 **Tiffany had pledged not to source gold from Pebble Mine should it be built:** Michael J. Kowalski, the CEO, has this quote on the Tiffany and Co. Web site: "There are some special places where mining clearly does not represent the best long-term use of resources. In Bristol Bay, we believe the extraordinary salmon fishery clearly provides the best opportunity to benefit southwestern Alaskan communities in a sustainable way. For Tiffany and Co.—and we believe for many of our fellow retail jewelers—this means we must look to other places to responsibly source our gold." http://www.tiffany.com/CSR/Common/CSRPrint.aspx?mode=page&key=textCSRSourcingPreservation.

204 **I recall Kass saying:** This is my own recollection of the exchange with Sam Kass. Pete Andrew and Shoren Brown recall things similarly. Reached later for verification of his remarks, Kass wrote, "I don't recall the specifics of what I said so cannot confirm or deny."

208 **"I'm not probably what you think about":** The event at Yale took place on January 18, 2012, and was recorded and transcribed by Kendall Barbery, master of environmental science candidate 2013, Yale School of Forestry and Environmental Studies, in May 2012.

211 **As Bill Carter recounts:** Bill Carter, *Red Summer: The Danger, Madness, and Exaltation of Salmon Fishing in a Remote Alaskan Village* (New York: Scribner, 2008).

214 **salmon prices fell precipitously:** Salmon aquaculture rose from
12,800 metric tons in 1980 to 1.3 million metric tons in 2001—
roughly 100 times more fish farmed in twenty years. Norway
topped the United States as a salmon supplier in 1997. Chilean
farms outproduced the United States wild and farmed by 2000. As
a result of the expanding supply of imported farmed salmon, the
prices fishermen were getting for wild Alaskan sockeye, chinook,
and coho salmon dropped an average of $1.30 per kg of fish from
1990 to 2001. For the whole Alaskan salmon harvest, the ex-vessel
price (or price the fishermen get, not market price) dropped to less
than half, from $570.2 million in 1990 to $216 million in 2001.
These statistics from Trond Bjørndal, Gunnar A. Knapp, and
Audun Lem, "Salmon: A Study of Global Supply and Demand,"
Food and Agriculture Organization of the United Nations, Globe-
fish, Fishery Industries Division, July 2003.

214 **suffered at least a 30 percent loss in income:** Fishermen in the herring industry closest to the spill lost everything—and the fish
didn't come back. Many fishermen went bankrupt. See Dan Simon
and Augie Martin, "Alaska Fishermen Still Struggling 21 Years
After Exxon Spill," CNN U.S., May 7, 2010, http://www.cnn
.com/2010/US/05/06/exxon.valdez.alaska/.

216 **coupled with a regulator protein from an ocean pout:** Curiously,
the ocean pout is not kosher. I wondered if this would render the
AquaBounty salmon *trayf*, i.e. nonkosher. But when I asked Rabbi
Menachem Genack, the CEO of the Orthodox Union Kosher Division, he told me that a fish's kosher qualities depend primarily on
its physical characteristics. A kosher fish must have fins and scales,
and the AquaBounty fish has these external traits.

217 **If Alaska continued as America's premier fishing place rather
than becoming a resource extraction place:** NOAA reported that
Americans consumed about 4.9 billion pounds of seafood in 2009
and 2010 (www.noaanews.noaa.gov/stories2011/20110907_usfish
eriesreport.html) while the Alaskan catch of fish and shellfish in
2011 was 5.35 billion pounds—Alaska catches enough for American consumption with over a quarter of a billion pounds to spare
(www.akrdc.org/issues/fisheries/overview.html).

217 **AquaBounty's Stotish:** A full transcript of the proceedings of this
session is available at www.gpo.gov/fdsys/pkg/CHRG-112shrg
78022/html/CHRG-112shrg78022.htm.

219 **and the production of those farms could increase by two hundred
million pounds:** The United States produces about 20,000 tons, or

4 million pounds, of farmed salmon (www.fishwatch.gov/seafood _profiles/species/salmon/species_pages/atlantic_salmon_farmed .htm). Canada produces about 100,000 metric tons, or 220 million pounds (www.aquaculture.ca/files/documents/AquacultureStatisti cs2011.pdf), making the North American production approximately 224 million pounds. If AquAdvantage Salmon grow twice as fast, that's twice the production, so there would be an additional 200 million pounds.

224 **"severe impacts on salmon and detrimental, long-term impacts on salmon habitat":** This section of the first review draft was quoted in a *Washington Post* article by Juliet Eilperin, "Bristol Bay Mining Would Harm Alaska Salmon Habitat, EPA Analysis Says," *Washington Post*, May 18, 2012, http://articles.washingtonpost .com/2012-05-18/national/35457451_1_northern-dynasty -minerals-pebble-mine-project-gold-and-copper-mine. Since an extended public review period, a second review draft with similar risk conclusions has been released and is now open for public comment; it can be found here: http://cfpub.epa.gov/ncea/bristolbay/ recordisplay.cfm?deid=242810#Download.

224 **Anglo American, the British mining giant, decided to withdraw from the Pebble project:** For more on the story, see Brad Wieners, "Why Miners Walked Away from the Planet's Richest Undeveloped Gold Deposit," *BloombergBusinessweek*, September 27, 2013, www.businessweek.com/articles/2013-09-27/why-anglo-ameri can-walked-away-from-the-pebble-mine-gold-deposit.

CONCLUSION

233 **minuscule in comparison with the 202 pounds:** The latest available figures for total meat consumption are through 2012, with the estimates for 2013 and 2014 from the National Chicken Council based on USDA statistics: www.nationalchickencouncil.org/ about-the-industry/statistics/per-capita-consumption-of-poultry -and-livestock-1965-to-estimated-2012-in-pounds/.

235 **over eighty thousand large dams:** From the National Inventory of Dams conducted by the U.S. Army Corps of Engineers. The full statistic is 84,000 dams in the United States as of 2010 that are over twenty-five feet, hold more than fifty acre-feet of water, or "are considered a significant hazard if they fail." Go to: www.agc

.army.mil/Media/FactSheets/FactSheetArticleView/tabid/11913/
Article/10236/national-inventory-of-dams.aspx.

236 **Overfishing has greatly decreased in American waters:** A summary of the successes of Magnuson-Stevens can be found in the
Natural Resources Defense Council report: Brad Sewell, "Bringing Back the Fish: An Evaluation of U.S. Fisheries Rebuilding
Under the Magnuson-Stevens Fishery Conservation and Management Act," Natural Resources Defense Council, February 2013,
www.nrdc.org/oceans/rebuilding-fisheries.asp.

236 **This in turn causes more fishermen:** A North Carolina study
found that about a third of local fish houses had gone out of business due to the combination of real estate speculation and low seafood prices. See Anne S. Deaton et al., "North Carolina Coastal
Habitat Protection Plan," North Carolina Departemnt of Environment and Natural Resources, October 2010, www.ncleg.net/
DocumentSites/Committees/JLCSA/Reports%20Received/
2010-2011/Dept%20of%20Environment%20and%20Natural%
20Resources/2010-Sept%20-%20CHPP%20Draft%20Report
.pdf.

238 **top six seafoods the average American consumes:** See NOAA's
FishWatch page on the top ten consumed seafoods in the United
States: www.fishwatch.gov/features/top10seafoods_and_sources
_10_10_12.html.

240 **the New Amsterdam Market, a small group of local citizens:**
Founded in 2005 by Robert LaValva, the New Amsterdam Market
has grown into a community of vendors from in or near New York
City that produce responsible agriculture, fish, and other food
products. Six seasons of markets have been held at the Old Fulton
Fish Market site throughout the year. To learn more, visit http://
www.newamsterdammarket.org.

241 **Howard Hughes Development Corporation:** Excellent reports on
Howard Hughes's activities in the South Street Seaport can be
found in New York's *Downtown Express* by the dedicated reporter
Tereze Loeb Kruezer, "Seeking a View into the Seaport's Future,"
Downtown Express, September 25, 2013, www.downtownexpress
.com/2013/09/25/seeking-a-view-into-the-seaports-future/ and
Tereze Loeb Kruezer and Josh Rogers, "Seaport Tower Plan Is
Met with Boos," *Downtown Express*, November 20 and 21, 2013,
http://www.downtownexpress.com/2013/11/20/angry-reaction
-to-plan-for-seaport-tower-marina/.

242   **Maine-based oyster and mussel grower:** The grower was Carter Newell of the Pemaquid Oyster Company. http://www.pemaquidoysters.com.

244   **"Gulf Wild" tag:** Gulf Wild is run by the Gulf of Mexico Reef Fish Shareholder's Alliance, a nonprofit trade association that partnered with the Community Seafood Initiative, State of Florida Department of Agriculture, the Environmental Defense Fund, and others to develop the tagging program. More information on Gulf Wild can be found here: http://mygulfwild.com/.

246   **a new Montauk-based venture called Dock to Dish:** Barrett and more than thirty-six commercial fishermen and wild shellfish harvesters from Suffolk County, Long Island, founded Dock to Dish. For more information, see http://docktodish.com.

247   **at least forty different community-supported fisheries have been born and are now active in nearly every American coastal state:** For CSFs in business as of the end of 2013, see LocalCatch.org's list at http://www.localcatch.org/about.html and Northwest Atlantic Marine Alliance's list at http://namanet.org/csf.

Index